D0932735

Hidden Knowledge
Organized Labour in the
Information Age

D.W. Livingstone
Peter H. Sawchuk

Garamond Press
Aurora, Ontario

© The authors, 2004

No part of this book may be reproduced or transmitted in any form, by any means, electronic or mechanical, without permission of the publisher, except by a reviewer who may quote brief passages in a review.

Printed and bound in Canada

Garamond Press Ltd,
63 Mahogany Court,
Aurora, Ontario L4G 6M8
www.garamond.ca

Cover photograph courtesy of Peter Sawchuk

National Library of Canada Cataloguing in Publication

Livingstone, D. W., 1943-
Hidden knowledge : organized labour in the information
age / D.W. Livingstone, Peter H. Sawchuk.

Includes bibliographical references and index.
ISBN 1-55193-045-5

1. Labor union members—Education—Canada. 2. Working
class—Education—Canada. 3. Labor union members—
Canada—Interviews. 4. Working class—Canada—Interviews.
5. Adult learning—Canada—Case studies. 6. Information
society—Canada. I. Sawchuk, Peter H. (Peter Harold), 1968-
II. Title.

LC5225.L42L59 2004   374'.0088'3317   C2003-905107-2

Garamond Press gratefully acknowledges the support of the Department
of Canadian Heritage, Government of Canada, and the support of the Ontario
Media Development Corporation of the Government of Ontario.

Contents

Tables

Figures

Acknowledgements

This book had its origins in David's early experience in road construction and food processing plant union locals in British Columbia; followed later by involvement with an inner city community school project in Toronto and research with one of the largest steel union locals in Canada (Seccombe and Livingstone, 1999; Livingstone and Smith forthcoming). These were all creative working-class organizations whose collective knowledge and skills were given little credibility by employers and other authorities. The book is also rooted in Peter's experiences working as a press operator in an auto parts manufacturing plant (see Sawchuk, 2003) as well as his involvement with unemployed workers' coalitions and unions, including the Communications, Energy and Paperworkers Union and the United Steelworkers of America. The current study builds directly on these prior work experiences.

The current project would never have got off the ground without the guidance of the experienced labour educators who acted as co-investigators from the outset. The combined talents and insights of Mike Hersh, D'Arcy Martin and Jennifer Stephen provided an essential foundation for our links with the specific unions that agreed to participate. Little of the actual research would have been possible without the support and substantial donation of time and energy by the unions and union locals involved, namely the Canadian Auto Workers (CAW); the Ontario Region of the Communication, Energy and Paperworkers (CEP); the Ontario Public Service Employees Union (OPSEU); District 6 of the United Steelworkers of America (USWA); the International Ladies Garment Workers Union (ILGWU); and the Union of Needletrades, Industrial and Textile Employees (UNITE). Individual labour leaders who must be mentioned as key figures in the research process include John Kovac, Mike Shields and Lyn Brophy (CAW); Tam Gallagher and Dave Zidenberg (CEP); Janice Hagan and Eugene Wilson (OPSEU); Mike Seward, Dan DeLeon, Jorge Garcia Orgales and Andy King (USWA); Alex Dagg and Jan Burowoy (ILGWU); Jonathan Eaton, Barrie Fowlie and Pat Sullivan (UNITE);

Mike Seward, past president of the Toronto District Labour Council and Jim Turk, former education director of the Ontario Federation of Labour. We are especially grateful to the key informants and interviewed workers at all sites who kindly shared their views with us. We hope they will find their learning experiences reflected truthfully in the text.

The Working Class Learning Strategies Project was funded by the Social Sciences and Research Council of Canada (SSHRC Grant #884-94-0018) from 1995 to 1998. Research on the family and community aspects of the study continued under the Working Class Learning Practices Project, also funded by the SSHRC under the national research network on New Approaches to Lifelong Learning (NALL) from 1998 to 2002 (see <www.nall.ca>). The Ontario Institute for Studies in Education of the University of Toronto (OISE/UT) provided graduate student support. The Department of Sociology and Equity Studies (SESE) at OISE/UT served as the operations centre for the field studies. The new Centre for the Study of Education and Work (CSEW), sponsored by SESE and the Department of Adult Education and Counselling Psychology at OISE/UT, grew directly out of the work on these projects and is continuing related research. Many OISE colleagues have offered helpful advice, including Kari Dehli, Kiran Mirchandani, Roxana Ng, Jack Quarter, Wally Seccombe and Dorothy Smith.

The core research team included D'Arcy Martin who, in addition to his co-investigator role, wrote initial drafts of the research process chapter and later took on the role of research coordinator of CSEW. Reuben Roth did all the interviewing at CAW Local 222 and, as a former assembly worker in the auto plant, provided invaluable assistance in the research design. He wrote the initial site report and co-wrote the auto chapter, conducted his MA and PhD theses drawing on this local and also served as research coordinator of NALL. Clara Morgan conducted some of the garment worker interviews, wrote the preliminary sector report, assisted with the garment chapter, and also completed a related MA thesis. Megan Terepocki did many initial interviews at the small parts manufacturing and college sites as well as preliminary report work, prior to going to work with the BC labour movement in the middle of the project. Rachel Gorman took over some of Megan's responsibilities at the college site and also began a related site study with disabled workers (see Gorman, 2002) which, unfortunately, was not completed because the group disbanded. Darlene Clover also provided research assistance with the college and garment reports. Additional interviewing was carried out by a talented group of student researchers, including Pramila Agrawal, Robert Bowd, Marnina

Gonick, Lorena Gajardo, Lisa Gribowski, Josephine Mazucca, Guida Man, Laura Mitchell, Paul Raun and Renita Wong. Doug Hart contributed assistance with quantitative analyses, and Milosh Raykov, Stephanie Percival and Fab Antonelli conducted literature reviews.

The project was designed and carried out as a form of participatory research in solidarity with working people and their organizations, but we take responsibility for any errors, omissions or shortcomings of the book.

Our greatest debts are to Angela and Jill, our partners, who gave continual love and support in spite of many lost weekends.

Our dedication for the book is expressed across different generations. During the final stages of writing, Peter's grandfather, Harry Sawchuk — a Ukrainian immigrant without any formal education, a nickel processor and a Mine Mill union member — passed away. Around the same time, David's grandchildren, Brooke and Leah, were born. It is our hope that the hidden knowledge of the previous generation of workers will be more justly recognized and rewarded amongst the next.

D.W. Livingstone
Peter H. Sawchuk
August 2003, Toronto, Canada

Dimensions of Learning and Work in the Knowledge Society

Working people are more knowledgeable and actively engaged in learning than public discussion generally assumes. Two basic assumptions underlie much recent discussion about work and learning: 1) a new "knowledge-based economy" is quickly emerging with new jobs generally requiring greater knowledge and skill; and 2) a "lifelong learning culture" must be created in order for workers to cope with these employment-related knowledge demands. Virtually every recent public policy statement about employment in every advanced industrial country begins with these assumptions. It implies that most workers suffer from a deficit of necessary skills and knowledge which must be rectified by greater education and training efforts.[1] We challenge these assumptions.

Constant change has been the hallmark of industrial capitalist societies since their inception in the early 1800s. Paid workplaces in capitalist economies are characterized by continual change as enterprises compete with each other for commodity markets, employers negotiate with employees over maximizing profits and/or creating fulfilling working conditions, and new technologies are adopted to aid all of these objectives. In the twentieth century, capital intensification in extractive and manufacturing industries has put an increasing premium on human mediation of expensive machinery. The rise of the service sector has been contingent on the selling of labour-intensive services rather than material goods. The proliferation of information technologies has made a wider array of work tasks dependent on workers' initiative. In short, there has been a secular trend in which the motives and learning capacities of the workforce play a more central, strategic role in capitalist labour processes. The dominant discourse of management theory has shifted from advocacy of scientific management of workers' bodily movements through extrinsic rewards to

promotion of learning organizations designed to enable continuing learning and enhance worker motivation to share their knowledge (see Boud
and Garrick, 1999; Bratton, Helms-Mills, Pyrch and Sawchuk, 2003).
While actual working conditions in most paid workplaces may seriously
diverge from idealized versions of such "learning organizations," there is
little doubt that employers, employees and researchers alike are paying
more concerted attention to workplace learning activities as we enter the
twenty-first century (e.g., Belanger and Federighi, 2000; Garrick, 1999;
and Lasch and Urry, 1994).

In this context, the dominant discourse incessantly repeats the mantra
that a "knowledge-based economy" requires a much higher proportion of
highly skilled workers and the creation of a "learning society" is imperative for people to acquire the additional knowledge and skills needed to
survive in this new economy (see OECD, 1998: 10). However, there is
considerable empirical evidence that the reality may actually be the reverse of this. That is, careful assessments of the changing occupational
composition of the employed labour force and of specific vocational
preparation requirements for the aggregate array of jobs in countries like
Canada and the United States have found only very gradual net upgrading
of the actual skill requirements of jobs over the past few generations
(Barton, 2000; Handel, 2000; Lavoie and Roy, 1998; Leckie, 1996). On
the other hand, rates of completion of post-compulsory schooling and participation in further education courses have grown exponentially during
the same period (Livingstone, 2002).The increases in educational attainment appear to be outpacing increases in job requirements and suggest
that we may already live in a "learning society," but not yet in a "knowledge-based economy."

All truth claims are subject to the interplay of knowledge forms and
power relations. Dominant discourses or ideologies about work and learning tend to reflect the interests of the most powerful economic groups.
Critical social theorists working against dominant forms have helped to
deconstruct such regimes of truth, as in Foucault's (1977) documenting of
the historical construction of coercive institutions, and in Herman and
Chomsky's (1988) analysis of the ideological effects of mass media on
popular opinion. But critiques of dominant social forms, in and of themselves, shed little light on the actual work and learning practices of most
people; that is, the practices that are most directly constitutive of everyday
life (e.g., de Certeau, 1984). This is not to suggest that there exist some
natural popular standpoints amongst ordinary people that are unaffected
by dominant ideologies of work, learning or social life generally, quite the
contrary (see Rose, 1990). But the less powerful mediate dominant truth

claims through their own direct experience and both their perceptions and their practices may well differ from articulated dominant versions of social reality (e.g., Seccombe and Livingstone, 1999). Empirical studies conducted from the standpoints of ordinary working people (e.g., Howell, 1973; Luttrell, 1997; MacLeod, 1995; Sawchuk, 2003a; Sennett and Cobb, 1972) are a necessary supplement to dominant discourses and scholarly critiques in order to comprehend the contemporary character of work and learning. This book will suggest a more inclusive conceptual framework for studying multiple dimensions of learning, work and their interrelations, present illustrative findings from our case studies of learning and work practices, and suggest some implications for further research and practice.

Our research starts by questioning the dominant discourse on learning and knowledge, and proceeds to document the actual activities of working people. Our evidence, based mainly on a series of in-depth interviews with unionized workers and their families, presents a picture that diverges sharply from dominant stereotypes to reveal highly active learners who face serious barriers to applying much of their current skill and knowledge in their paid workplaces, formal educational settings and civil society generally. In fact, working people are far more likely to be underemployed in their jobs than to be underqualified for them. A growing body of survey-based studies has confirmed this condition in many countries.[2]

But this reality is difficult for many people to accept because they continue to hear knowledge deficit claims (see Valencia, 1997) asserted so frequently by opinion leaders – who tend to focus on chronic skilled trades shortages and occasional shortages of specific professional and technical workers, which are exceptions rather than the general rule. Instead of recognizing the rich skill and knowledge base of the general population that could serve to overcome these specific shortages through, for example, prior learning assessment and recognition, the fixation with these specific deficits is typically taken to demonstrate general shortages. This book looks beyond both the dominant rhetoric of skill deficits and the growing survey data on underemployment to the direct experiences of working people.

Knowledge and power are intimately related. The most powerful people – corporate executives, top managers, and professionals – are most likely to have their knowledge and skills institutionally certified and closely linked with opportunities to apply these capabilities. The least powerful people – those hired for an hourly wage and the unemployed – are most likely to have their knowledge and skills institutionally ignored or devalued and to have only more fragmented and submerged chances to apply their capabilities in most paid workplaces or other public settings (Sawchuk, 2003a). Through our focus on organized wage workers, we have, in a sense, chosen to do our research with the *most* powerful among the *least* powerful. In fact, working people, to contest the exploitation of their labour power as well as the appropriation, degradation or suppression of their skills and knowledge, have often banded together in labour unions and related social movements based in their paid workplaces. It is within and because of these organizations that working people have been most able to share information, develop common world views, traditions and strategies, and mobilize capacities towards their own collective interests. It is through these organizations that "working-class cultures," indigenous knowledge and distinctive learning practices of working people (Chapter 2) are likely to be most visible and accessible. Our focus is therefore on union locals as primary sites of selection for this study. There are substantial variations among these unions in economic power and consequent educational provisions for their members, as we shall see. But we will also see that *informal learning* holds a special place for subordinate social groups such as these, and that it is much more extensive amongst all unionized workers than any type of formal education. At the same time, the organized core of working people represented in this study should not be easily assumed to be more knowledgeable and skilled than the unorganized periphery just because they are relatively more powerful – any more than corporate executives should be assumed to be more intelligent than union leaders just because they have more economic power. A similar argument can be applied to the capabilities of women homemakers, who are an unorganised labour force and whose skills remain hidden in the household. Innate intelligence and acquired skills and knowledge vary widely in all social positions, indeed much more widely *within* social groups than *between* them (Curtis, Livingstone and Smaller, 1992). It is a self-serving conceit of the powerful that they are necessarily more intelligent and skilled than the powerless. Finally, none of these observations is intended to denigrate formal educational attainments, but rather to suggest that the formal attainments of working people are currently underutilised in paid workplaces and their informal learning capacities are even more widely ignored.

Spheres of Work and Learning

Our study begins by selecting organized workers from paid workplace sites, partly in terms of varied levels of formal schooling. But we intend to go beyond conventional thinking and research about work and learning which generally suffers from narrow conceptions of both phenomena. In economically advanced societies, there are at least three distinguishable spheres of work – paid employment, housework and community volunteer work – and four spheres of learning – initial formal schooling, further or adult education, informal training and self-directed informal learning.

"Work" is commonly regarded as synonymous with "earning a living" through *paid* (or, more rarely, unpaid) *employment* in the production, distribution and exchange of goods and services commodities. Such presumptions conflate the "job-place" with the "work-place" (Bratton *et al.*, 2003). While we will analyze paid employment specifically, we will also examine other important forms of work. Most of us must do some household work, and many of us need to contribute to community labours in order to reproduce ourselves and society. Both *housework* and *community volunteer work* are typically unpaid and underappreciated, but they remain essential for our survival and quality of life (see Waring, 1988). Furthermore, the relations between paid work, housework and community work may represent major dimensions of future economic change. Men and women continue to renegotiate household divisions of labour, while more and more aspects of housework and community work are being transformed into new forms of paid employment.

In its most generic sense, *learning* involves the acquisition of understanding, knowledge or skills, anytime and anywhere. It takes place throughout our lives, and the sites where it occurs make up a continuum, ranging from spontaneous and tacit responses to everyday life to highly organized participation in formal education programs. Formal schooling, adult education and informal learning are the three forms of intentional learning that researchers now commonly identify. *Formal schooling* is a sequentially structured and hierarchical series of curricula and credentialing programs of study typically administered at elementary, secondary and tertiary levels. It happens in settings organized by institutional authorities, planned and directed by teachers approved by these authorities, and typically requires compulsory attendance until mid-adolescence. *Adult education* includes a diverse array of further education programs, courses and workshops with authorized instructors in many institutionally organized settings, from schools to workplaces and community centres, and it is typically voluntary. Adult education is the most evident form of lifelong learning for adults past the initial cycle of schooling. But people also continually engage in informal learning activities to acquire understanding, knowledge or skills outside of the curricula of institutions providing educational programs, courses or workshops. Informal learning – which we undertake individually or collectively without externally imposed criteria or the presence of an institutionally authorized instructor –

is much more widespread amongst adults than either initial school attendance or further adult education. Informal learning includes both *informal training* provided by more experienced mentors and *self-directed learning* which we do on our own or with peers.[3] It displays many self-conscious as well as tacit dimensions, the latter of which are only now being carefully explored (e.g., Sawchuk, 2003b). As Allen Tough (1978) has observed, informal learning is the submerged part of the "iceberg" of adult learning activities. It is at least arguable that, for most adults, informal learning represents our most important form of learning for coping with our changing environment. Beneath the surface of formal schooling and further or adult education, no account of a person's "lifelong learning" is complete without considering their informal learning activities.

In sum, both work and learning are more extensive and complex phenomena than is often implied in discussions of employment and education. A narrow focus on relations between paid employment and organized education ignores significant interrelations between these other dimensions of work and learning as well as interrelations across broad spheres of activity including home and community practices. Early informal childhood socialization is increasingly recognized as highly influential in determining success in formal schooling. There is far less appreciation of the fact that continued informal learning is vital for success in paid workplaces. Recent studies confirm that most job-related training is done informally (see Betcherman *et al.*, 1997; Center for Workforce Development, 1998). The majority of workers manage to become at least adequately qualified for their jobs through a combination of initial schooling, further adult education, informal training and self-directed learning. Even so, the dominant discourse about a pressing need for the creation of "learning organizations" ignores these realities of interaction between organized education, informal learning and job performance, presuming instead that the central challenge for improved enterprise performance is for workers to become more active and motivated learners. Also, many valuable transfers of knowledge and skill between these four basic spheres of learning and among the three spheres of work are similarly unrecognized or discouraged by actual workplace organization (see Livingstone, 1999a; Sawchuk, 2003a).

Another disadvantage in much research about paid work and education is that most studies focus too narrowly on immediate payoffs to employers. From a short-term management perspective, virtually the only relevant learning for employees is job training that quickly enhances the company's productivity or profitability on a quarterly basis. From this vantage point, much learning that workers gain both on and off the job is effectively non-existent. However, preliminary studies from workers' standpoints have discovered, for example, that many assembly-line workers develop informal learning networks to learn how to use personal computers. Some become competent computer programmers, despite having no employer encouragement or immediate formal opportunities to use these skills in their jobs (e.g., Sawchuk, 1996, 2003a). What workers learn informally on and off the job is at least potentially applicable both in jobs redesigned to use workers' growing repertoire of skills more fully, and in other socially useful and fulfilling household and community activities. We need to find out how relevant this more general and informal knowledge is, and not continue to ignore it. This study attempts to address this gap directly by proceeding from workers' own standpoints (Chapter 1).

Taking into account unpaid work and informal learning generates much more extensive general profiles of the actual activity patterns of adults. For example, the total adult population in most advanced industrial countries is now spending as much time in unpaid household and community work as in paid employment, averaging around fifty hours of work per week (Statistics Canada, 1999b). People employed full-time as well as homemakers with children at home currently work around sixty hours per week in all forms of paid and unpaid work. Those not in the employed labour force generally do quite a lot of unpaid work, including over twenty hours per week of housework by unemployed and retired women and an average of around three hours per week of community work in all social groups. There are continual changes in employment conditions, including the growth of service sector occupations, an increase in part-time jobs and polarization of employment hours to produce both overwork and underemployment, and diffusion of information technology through paid workplaces. People continually move in and out of the official labour force; about a quarter of current women homemakers expect to return to employment within the next year (Livingstone, 2002).

In terms of learning, all spheres appear to have experienced rapid growth during the past generation. Participation in advanced schooling has increased exponentially. Between 1961 and 1998, the proportion of the Canadian 25-to-29 cohort who completed university degrees increased from 4 percent to 26 percent (Livingstone, 2002: 18). Canada now leads the world in its levels of post-secondary education with nearly half of the age 20–64 population having attained some form of post-secondary credential by 1996 (Statistics Canada, 2000). Adult course participation may have expanded even more rapidly. In Canada, according to government surveys, the participation rate grew from 4 percent in 1960 to 35 percent in the early 1990s (Livingstone, 2002: 20–24). There is less direct evidence on informal learning. The first national survey on this was only conducted in Canada in 1998 and the only prior comparable national survey was in the United States in 1976 (see Livingstone, 2002: 24–32). But, according to their self-reports in the 1998 survey, around 95 percent of Canadian adults were devoting some time to *intentional* informal learning activities related to their paid employment, household duties, community volunteer work and other general interests, an average of about fifteen hours per week (Livingstone, 1999b). So, the incidence of intentional informal learning also appears to have increased substantially since the first empirical studies conducted in the 1960s and 1970s estimated averages of around ten hours per week (Tough, 1978). Those in the employed labour force now report spending an average of six hours per week in job-related informal training and non-taught learning pursuits. The participation rates and time involved in informal learning are much greater than in adult education courses in which only around a third of all adults currently spend an average of only a few hours per week (Doray and Arrowsmith, 2001; Statistics Canada, 2000). The suggestion that adult learning is like an iceberg, with most of it submerged informal learning, appears to be very appropriate. More generally, the findings suggest that, by any reasonable definition, Canada and, probably, most other advanced industrial countries are already "learning societies" or "knowledge societies."

Class and Learning

It is also rare in dominant discourses to hear any reference to social classes. While much attention is paid to the exploits of the rich and famous and much less to the plight of the homeless, the rest of us are generally assumed to be "middle class." Increasing income polarization makes this image more difficult to sustain, but the underlying reality is that there has always been a strong class structure in capitalist societies, generated by paid workplace relations and linked closely with activities in many other aspects of social life. Social classes are relational rather than categorical in character so simple classification schemes are not very informative. But it is important for a class analysis such as ours to be as specific as possible about major active class groupings (see Livingstone and Mangan, 1996). *Corporate capitalists* who own large assets and employ many others can live very affluently. Other proprietorial classes include *small employers* who own their own enterprises with fewer employees and the *self-employed* who survive through businesses based on their own labour. The larger private enterprises and public organizations hire *managers* to control their regular operations and *professional employees* to perform various specialized functions semi-autonomously. In larger or more dispersed organizations, managers delegate some operational control roles to supervisors and forepersons. The rest of the employed workforce who do not own their organizations, have no official authority roles and no autonomous control over work processes are the *working-class*; *industrial workers* who produce goods and *service workers* who provide various clerical and sales services. In this book, the term "working-class" generally refers to industrial and service workers. Outside the actively employed labour force, there is an underclass or lumpen class made up of a variety of people unable to find sustaining employment. There are also others who are connected to the class structure of the active labour force through their personal trajectories or household relations, including *students, retired people* and *homemakers*. As wages and salaries have become the increasingly pervasive form of labour and enterprises have become larger over the past century, more and more intermediate managerial and professional employee class positions between capitalist owners and the working-class have been created. The massive entry of married women into paid employment in the post–World War II era has produced many double-income households. As always, capitalist relations of production are continually changing, so the class structure is increasingly complex. But different class locations in these terms continue to have real consequences in terms of social consciousness, life chances and lifestyles, as we and others have extensively documented elsewhere (see Seccombe and Livingstone,1999; Wright, 1996).

The fact remains that the majority of those born into the working-class in most advanced industrial societies remain in the working-class throughout their lives (Livingstone and Stowe, 2003) and our primary interest here is their knowledge acquisition processes. We must begin by registering the well-documented finding that those from working-class origins have been persistently underrepresented in higher education. Not only do those from higher class origins have a much greater likelihood of obtaining post-secondary degrees, but those currently in higher class positions are also more likely to continue to participate in further education courses and workshops. As Table 0.1 shows, Canadian corporate executives, professional employees and managers are much more likely to have university degrees than most other occupational classes while service workers and industrial workers are least likely; corporate executives are about ten times more likely to have a university degree than are industrial workers. Adult education is somewhat more evenly distributed between occupational classes. Surveys based on narrow definitions of formal course participation typically find much greater enrolment rates for professional and managerial employees than for workers (Arrowsmith and Oikawa, 2001:

35). Even when workshops of very short duration are included, as in Table 0.1, corporate executives, managers and professional employees are about twice as likely as industrial workers to have participated in the past year. Further analysis finds that unionized workers, who often have more bargaining power, tend to have higher participation rates in, and employer support for, further education courses than do non-union workers (Livingstone, 2002; Sawchuk, in press).

Table 0.1: Occupational Class by Schooling, Further Education and Incidence of Informal Learning, Employed Labour Force, Canada, 1998

Occupational Class	University Degree	Course or Workshop Past Year (%)	Employment-related Informal Learning (%)	Total Informal Learning (Hrs/week)
Corporate executives	70	71	98	17
Small employers	22	52	97	16
Self-employed	15	52	91	14
Managers	34	73	90	13
Professionals	40	67	92	15
Supervisors	12	63	87	14
Service workers	8	54	81	17
Industrial workers	4	33	83	17
Total (N=951)	17	56	86	16

Source: Livingstone (2002).

But the second vital point to recognize is that the distribution of the incidence of self-reported informal learning appears to be *quite equitable* regardless of occupational class, prior schooling or adult education participation. The more highly educated occupational classes are only marginally more likely to be involved in employment-related informal learning and no more likely at all to spend time doing informal learning generally. Around 90 percent of those in all occupational classes indicate involvement in employment-related informal learning and the average time devoted to informal learning generally is around fifteen hours per week in all classes. Service workers and industrial workers are just as actively involved in learning activities that they control themselves as are occupational classes with greater economic power.

Further analyses of the interrelations between work time and learning time find that full-time workers are somewhat more likely than part-time workers to participate in further education courses but that the longer hours people are employed, the less time they tend to spend on job-related courses. However, there are generally positive associations between the amount of time that people spend in paid employment, household labours and community work and the time they spend in the respective types of work-related informal learning. The association between community volunteer work and community-work-related informal learning is much stronger than the relation between paid employment and job-related informal learning (see Livingstone, 2002: 38–40). These findings suggest that the greater degree of discretionary control one has to engage in the particular sphere of work, the closer the relation between work time and informal learning time. These findings are compatible with earlier research that discovered reciprocal effects between holding less supervised jobs and engaging in more intellectually demanding "leisure-time activities" such as hobbies and general interest reading (see Kohn and Schooler, 1983). But the survey results suggest that informal learning is now also pursued extensively by those holding more routine, highly supervised working-class jobs. Such research has probably only touched the tip of the iceberg of adult learning, especially the learning of working-class people. Our study tries to go deeper into this largely hidden dimension of working-class learning

Working-Class Underemployment

Prior studies revealed that members of the active labour force have achieved rapid increases in their formal educational attainments and adult education participation rates, while also pursuing vast and increasing amounts of informal adult learning. By most reputable measures, the skill and knowledge requirements of the job structure have experienced much slower growth. It follows that, in overall terms, the cumulative employment-related knowledge and skills of the potential labour force probably now exceed the capacity of the current labour market to provide adequate numbers of corresponding jobs, a condition called *underemployment*.

For a comprehensive discussion of the multiple dimensions of underemployment see, for example, Livingstone, (1999a). Commonly recognized dimensions include: *structural unemployment, involuntary temporary employment, credential underemployment; performance underemployment* and *subjective underemployment*. The most relevant measures for our purposes focus on the employed labour force. Credential underemployment includes job holders who have attained at least one credential higher than is required for job entry; performance underemployment includes job holders whose achieved levels of skills and knowledge significantly exceed the levels required to do their job, regardless of any entry credentials; and subjective underemployment includes those whose self-assessment is that they are overqualified for the jobs they have held. Many Canadian studies have documented the general existence of some of these dimensions of underemployment since the 1960s (e.g., Statistics Canada, 1999c; Tandan, 1969). The most recent estimates indicate that about 20 percent of the employed labour force now consider themselves to be subjectively underemployed, while around 30 percent have at least one credential higher than required for entry to their current job, and measures based on the general educational level required to perform the job suggest that as much as half of the labour force may have skill levels that exceed those actually required (see Livingstone, 2002).

Analyses of these underemployment measures by occupational class provides further insight into current employment-education matching for the employed labour force. The basic patterns are summarized in Table 0.2.

Table 0.2: Incidence of Underemployment by Occupational Class, Employed Labour Force, Canada, 1998

Occupational group	Self-assessment (%)	Credential gap* (%)	Performance gap (%)
Corporate executives*	2	9	17
Small employers	16	28	47
Self-employed	14	30	43
Managers	22	13	31
Professionals	9	16	38
Supervisors	20	30	70
Service workers	30	41	77
Industrial workers	20	38	47
Totals	20	28	53

Source: Livingstone (2002). N=951.
*Data for Ontario from Livingstone, Hart and Davie (1999c).

The most consistent finding on all three measures is that corporate executives, who wield the most economic power, are least likely to be underemployed. Few consider themselves overqualified for their jobs and their incidence of underemployment on other measures is significantly lower than other occupational groupings. At the other extreme, service workers, including mainly clerical and sales workers, appear to have the highest levels of underemployment on all job-specific measures. Nearly one-third of service workers report feeling overqualified for their jobs, about 40 percent hold a higher credential than the job currently requires for entry, and on a general educational development (GED) based performance measure around three-quarters have more formal education than actually needed to do their job. All occupational groups have lower levels of underemployment on self-assessment and self-reported credential criteria than on independent performance measures. There may be a number of group-specific reasons for these discrepancies.

For example, the proprietorial classes, including corporate executives, small employers and the self-employed are unlikely to perceive limits on the use of their skills since they can set their own working conditions. However, the performance gap measures are based on their much more diverse occupational categories rather than their proprietorial status. Managers, at the top end of the authority structure among waged and salaried employees, tend to have consistent levels of underemployment on all measures and relatively low levels on the more independent measures. While professionals tend to rely most strongly and directly on their high formal educational credentials for job entry and have accordingly low levels of subjective and credential underemployment, they do tend to have less control than managers do over the application of their skills in actual working conditions. Supervisors (mainly drawn from the working-class), service workers and industrial workers, at the bottom of the occupational status hierarchy, have the highest levels of underemployment on actual performance measures and the greatest discrepancies between self-rated and independent measures. This difference may reflect their more limited discretion to actually use the credentialed skills they needed for their jobs, but the overall orderings of results on each of these measures by occupational group are generally consistent with the prediction that those in lower positions in terms of economic power are more likely to be underemployed. Therefore, in spite of their relatively low levels of formal schooling, those in working-class jobs are still less likely than those in dominant class positions to be able to use their formal schooling on the job. The finding that working-class employees are just as likely as others to engage in informal learning activities – which are usually unrecognized and unrewarded by employers – only serves to accentuate the likely extent of working-class underemployment. Again, we will explore this in greater depth.

Most of those in the active labour force are engaged in a wide array of continuing learning activities related to their current or prospective jobs.

This pursuit of additional knowledge, skill and understanding related to employment applies across different employment statuses and occupational classes. Extensive engagement in job-related learning even applies to the considerable numbers of the "underemployed," who already have much more knowledge and skill than their jobs require. Working-class employees appear to be more likely to be underemployed than more dominant economic classes but no less likely to continue to engage in extensive informal learning activities.

Information Access in the Computer Era

The end of the twentieth century is increasingly seen as the onset of an "information age," largely due to the increasing proliferation of information technologies (IT) that purport to provide quicker and easier access to diverse forms of data, information and knowledge. The dissemination of IT in the form of personal computers and the Internet has been extraordinary in recent years. While less than 20 percent of Canadian homes owned a personal computer in 1989 (Lowe, 1992: 83), by 1997 the proportion had more than doubled and continued to grow rapidly from 45 percent of homes in 1998 to 55 percent in 2000 (Sciadas, 2002). It has been a mere decade since a publicly accessible electronic information exchange network, the Internet, was created but by 2001, 60 percent of all households had at least one member who used the Internet regularly either from home or another location (Statistics Canada, 2002). Most North American adults – including working-class adults – now have access to computers and the Internet (see Weis, 2001).

The impact of new technologies on knowledge acquisition has typically been wildly exaggerated (see Cuban, 1986; Livingstone, 1997b) but the combination of personal computers and the Internet provide a more interactive and dynamic mode of acquiring knowledge than any previous form of IT. The vast majority of Internet users indicate that it has already had a significant impact on their lives, mostly by making them more knowledgeable by providing access to various information sources, and not by just making them more frequent shoppers (Angus Reid, 2000).

Despite significant growth in access and use of computers in the households of the employed working-class, the diffusion of home computers has been extremely uneven across economic groups, with recent ethnographic research (e.g., Schön, Sanyal and Mitchell, 1999) indicating important new exclusionary effects among low-income communities. Statistically, about three-quarters of the households in the highest income quintile had computers in 1998, compared with less than 20 percent of those in the lowest quintile (Statistics Canada, 13 December, 1999) and this gap continues to grow (Sciadas, 2002). This growing gap is the basis for justifiable social concerns about a "digital divide" among Canadians (Reddick, Boucher and Goseilliers, 2000) but general access and the capacity to use computers are much more widely distributed. Even in 1989, when less than a fifth of all households owned a computer, nearly half of the entire adult population was able to use a computer, and about one third had taken a computer course (Lowe, 1992: 71). A recent analysis based on in-depth interviews and ethnography analysis (Sawchuk 2003a) suggests that these basic access and use statistics tend to underrate working-class computer literacy activity by ignoring collective and informal dimensions of practice as well as exclusionary effects of dominant discourses of "learning." Nevertheless, both the diffusion of home computers and the development of basic computer literacy have continued to increase rapidly (Angus Reid, 2000).

According to most indications, adult workers have also continued to acquire computer skills to a greater extent than they have had opportunities to apply them in paid workplaces. The General Social Survey (GSS) in Canada indicates that, by 1989, around one third of the labour force was using computers for some tasks in paid workplaces. By 1994 that proportion had increased to 48 percent (Lowe, 1996). However, according to the GSS, in 1989 – when 35 percent of Canadian workers were actually using computers in their jobs – 59 percent of workers had the ability to perform work-related computer applications; by 1994, when 48 percent of all workers used computers in their jobs, computer literacy had increased to 68 percent of the employed workforce (Lowe, 2000: 75). Similarly, while around 80 percent of adults now have some form of Internet access, net users are far more likely to say that they use it to acquire general knowledge, for entertainment, personal communications and financial transactions than to improve their job performance (Angus Reid, 2000; Dickinson and Sciadas, 1999). We will examine this discrepancy between knowledge acquisition and use on the job more closely in our case studies of the service and industrial workers who appear to experience it most frequently.

We certainly appear to be living in a "knowledge society" in terms of the accessibility of information from multiple sources, and in a "learning society" in terms of the continuing learning efforts of most workers. Although extensive underemployment contradicts the frequent claims that we are also living in a "knowledge-based economy," the lack of immediate opportunities to use their new knowledge in available jobs does not appear to have dissuaded workers from continuing to seek ever more of it. There is considerable general evidence that we now have an active lifelong learning culture in the Canadian labour force, but one that seems to be insufficiently recognized in both dominant discourse and in many paid workplaces. With these general observations as a backdrop, we can now introduce our own site-based research findings, which are intended to shed further light on the specific processes involved in working-class learning from the standpoint of the working-class.

Sector-Based Research Sites

Employment organizations differ greatly in their general managerial practices and in their related approaches to worker training and use of worker knowledge. In comparative historical terms, increasing technical and social divisions of labour have required more planned coordination of labour processes and those organizations that have more effectively integrated the specialized knowledges of their workforces have tended to be most successful economically. Economic historians such as William Lazonick (1991) have suggested a progression of dominant business organizational forms over the past century, from the origins of "proprietary capitalism" in Britain through the "managerial capitalism" that emerged most distinctly in the United States beginning in the early 1900s, to the "collective capitalism" that can be most directly associated with Japanese firms at the close of the twentieth century. This progression can be understood pivotally in terms of the increasing extent of integration of the knowledge and skills of the labour forces. The organizations with the greatest capacity for innovation and quick response to environmental change are now generally considered to be those that invest most in developing employee skills and give operative workers most discretion in using them. The popular notion of the "learning organization" and knowledge-intensive work, while operative in niche markets for specific periods (e.g., Cohen and Sproull, 1996; Dixon, 1992; Frenkel, Korczynski, Shire and Tam, 1999), generally idealizes and exaggerates this tendency. But a central concern for management in dominant firms now involves attempts to consolidate and control knowledge production – that is, learning – and to gain privileged access to workers' creativity and capacities (through loyalty and commitment). Different organizational forms and managerial strategies can co-exist between sectors and firms, and even in some companies at the same time. It follows that different economic sectors, types of firms and organizational strategies of management may be related to different tendencies in worker-learning strategies.

Lazonick (1991) demonstrates the "myth of the market economy," specifically that leading firms tend to subvert open competition and free-market determination to achieve economic success. He shows that, in fact, the internalization, integration and cooperative coordination of production, distribution and exchange processes (within and between firms) leads to more advantageous capital accumulation and sustained economic survival and growth. Successful capitalist enterprises use these collaborative strategies as a means of obtaining privileged access to the "commodities" of human creativity and learning at below market value. Lazonick (1991: 88–89) summarizes the logic thus:

> When faced with the international challenge, the substantial productive resources that the dominant organization has at its disposal enables it to choose from two broad, and very distinct, competitive strategies – one innovative and the other adaptive. The innovative strategy is to plan, invest in, and create more powerful organizational and technological capabilities, perhaps coordinating the organization's strategy with privileged access to resources provided by the state. Alternatively, the adaptive strategy is to try to compete on the basis of productive capabilities inherited from the past.... In the face of competitors who are actively developing and utilizing their productive resources, the economic viability of the adaptive strategy may be prolonged by degrading product quality and by demanding longer harder work as well as pay concessions from employees. Depending on the extent of its prior competitive advantage, the productive capabilities of its competitors, the bargaining power and mobility of its employees, the quality requirements and brand loyalty of its customers, and the serviceability of its plant and equipment, the old leader may be able to make adequate profits for a period of time. In the long run, however, the once-dominant organization will eventually reach the limits of the adaptive strategy as it loses [privileged access to] productive employees and customers, as well as its productive capital base.

Such analyses as Lazonick's, while insightful in terms of organizational strategies, have been conducted largely from managerial standpoints and pay little comparative attention to differences in worker responses. The general strength and strategic responses of organized labour in unionized employment sites may be equally important to understanding organizational performance generally (cf. Mishel and Voos, 1992) and are certainly important mediating factors in comprehending the provision of employment-related learning opportunities for workers (e.g., Doeringer and Piore, 1971). Labour union approaches to promoting education and training programs may encourage worker learning even under the most regressive adaptive managerial regimes.

Our process of site selection for this research began with a concern to ensure that the sites represented a wide range of employment organizations and worker learning contexts. We therefore decided to focus on sectors of the economy with different employment trends and wage levels, innovative or adaptive managerial practices, union strength and training programs. As Table 0.3 summarizes, the basic industry profiles for the five sectors we chose differed quite widely on all of these features when we began our study in 1994–95.

Table 0.3: Organizational Dimensions of Employment Sites

Site	Wage Level*	Training	Managerial Practice	Employment	Union Strength
Auto	high	high	innovative/adaptive	down	very strong
Chemical	high	high	innovative	up	strong
College	medium/high	stratified	adaptive/innovative	down	moderate
Small parts	medium /low	low	adaptive	contingent	weak
Garment	low/medium	low	adaptive	contingent	very weak

*Note: "Wage Level" is defined relative to other sectoral sites in our study; "Training Policy" is defined by the existence of contract language directed specifically towards workers at the level of operative; "Managerial Practices" are based on policy statements by company, worker interviews and partial ethnographies; employment trends are based on industry statistics for the 1990–95 period; estimates of union strength are based on the subjective views of several independent observers.

The automotive industry had high established wage levels and fairly stable general employment levels. Union strength was significant with large concentrations of workers in a highly strategic industry. Managerial practices have been largely adaptive variations of neo-Taylorism through "lean production" but mixed with some innovative initiatives to nurture training in relation to new technology. An extensive array of worker education programs has been developed under pressure from a strong unionized workforce as well as increased global competition. The chemical industry was generally characterized by innovative managerial practices and relatively high commitment to worker training programs, as well as continuing employment growth, high wages and strong unions benefiting from cooperative strategies. Public sector service organizations have experienced pronounced employment reductions and intermediate wage levels. Managerial practices have been generally mixed with some innovative training policies focused mainly on managerial and professional employees. Union locals are typically of moderate strength bringing together fairly large numbers of workers located in diverse units. Small auto parts manufacturing firms have witnessed widely fluctuating employment patterns dependent on the needs of the larger firms they supply, with generally lower wages as well as more adaptive managerial practices and unions, where they existed, with quite limited bargaining power. There has consequently been little commitment to sustained training programs. Finally, the garment sector underwent major restructuring and fragmentation. Managerial practices were mainly adaptive and unions were seriously weakened by many factory closures and increasing reliance homeworking and sub-contracting. There were widely fluctuating employment levels and generally very low wage levels. Little commitment to work-based training programs was evident in the sector.

While we were committed to exploring a range of employment sites, the support of labour unions to provide effective access to a wide range of workers within each of these sectors was central in determining specific site selection. Our research project began with the active participation of three prominent labour educators (Mike Hersh, D'Arcy Martin and Jennifer Stephen) and relied heavily on their connections with specific unions and labour federations which then provided advice on the selection of appropriate employment sites and access to the leaders of union locals. The specific labour unions that provided support were:

· Automotive Sector: Canadian Auto Workers (*CAW*)
· Chemical Sector: Communication, Energy and Paper Workers (*CEP*)
· Public Service Sector: Ontario Public Service Employees Union (*OPSEU*)
· Small Parts Sector: United Steel Workers of America (*USWA*)
· Garment Sector: International Ladies Garment Workers Union (*ILGWU*) and United Needletrades, Industrial and Textile Employees (*UNITE*)

The actual method of the field research at each site is discussed in Chapter 1. It is most important to note that the specific sites reflect quite closely the general features of their respective sectors. Most pertinently, the stronger union locals in the higher wage sectors have typically been able to negotiate more extensive training programs and generally (though not exclusively) somewhat greater opportunity to develop and apply working-class knowledge forms.

In the automotive assembly plant, economic conditions are likely to produce relatively positive effects in terms of workers' learning practices. Industry Canada pointed out in its sectoral framework document during the period of our research (Automotive Industry Sector Competitive Framework, 1997) that the automotive assembly sector was in a "reasonable position," though it remained dependent on trade/tariff policy and the general economic environment which have now turned sour for workers in

Canada. But this mature sector, with a highly developed and innovative unionism (Livingstone and Roth, 1997) partially stabilized employment levels through the late 1980s and through much of the 1990s while production has steadily increased. There has been, for example, a large increase in unit production in the last few years, both across the industry and in the auto plant we researched. However, employment levels at this plant tell a different story. As the site chapter details, since 1984 – with sell-offs of a major section of the plant included – employment levels have dropped by almost 50 percent. Managerial practices have been "innovative" in the past partly because of organizational action by the union. It may be that the company has now adopted more adaptive strategies and thus, according to Lazonick's form of analysis, is "compet[ing] on the basis of productive capabilities inherited from the past" (1991: 88–89). In any case, through some combination of innovative and adaptive strategies, the plant, the firm and the entire auto assembly sector have continued to express a relatively strong commitment to worker training programs.

Workers at the chemical factory appear to be in a fairly advantageous position to engage in effective workplace learning. Their company is a relatively large, transnational market leader that consistently ranks in the top 200 net sales in the *Financial Post* indices and exhibits strong employment growth and strong capital investment. According to industry profiles, the chemical sector in Canada has weathered the 1980s and 1990s with relative ease, losing some jobs but maintaining strong overall growth in GDP, efficiency (GDP/employee) and capital investment (Industry Canada, 1997a). As Industry Canada notes, the sector is in fact poised for further growth should domestic conditions allow. According to Turner and Hadfield (1994), this growth has already been initiated by an invigorated worker-centred quality approach which stresses the centrality of continuous worker training.

The public sector union, a college staff union based at a community college, represents an organizational form that appears to have placed considerable stress on forms of innovative strategy (e.g., "Participatory Management Program;" Employee Development Centres and the "Worker-as-Educator" initiatives). But public sector economic fortunes are directly tied to the general political climate and fiscal policies of governments. With the rise of an ambitious right-wing provincial government in Ontario, financial conditions seemed to change almost overnight. As we document in the site chapter, between the 1995–96 and 1996–97 fiscal years alone, the total provincial funding to Ontario colleges dropped enormously. As workers at the site outline, suddenly the college was in a state of severe contraction, labour adjustment and "doing more with less." In this context, any innovative strategies the administration/management had entertained in the past gave way to adaptive ones focused on downsizing, work intensification, multi-tasking, "bumping" and reorganization issues. While such educational institutions are in the "business of training," both organizational commitment to staff training and the discretionary control by lower-level workers has generally tended to decline in such situations.

The small auto parts plant, our example of light manufacturing, is part of a privately-owned firm with several complementary units located in the same region, each with a mix of continuous run and batch production lines using largely outdated technologies. In this particular plant the company's investments are finely balanced, requiring it to make the most out of only one shift, overtime and work intensification mixed with periodic lay-offs and re-hiring based on (just-in-time) customer orders. Employment in the auto parts sector almost doubled between 1980 and 1995. As in the chemical industry, there was massive capital investment over the same period and steadily increased productivity (Industry Canada Report, 1997b). These conditions appeared relatively promising for innovative management and worker learning. However, there has since been a massive convergence, with large producers gobbling up smaller ones who because of increasing costs in the areas of technological upgrading, product development, greater quality assurance and warranty responsibility, can barely afford to operate. In addition, with auto parts manufacturers now operating globally, small Canadian firms increasingly face the threat of international suppliers undercutting their deals with the large automakers. Small regional parts suppliers, with little capital in comparison to the global giants of the sector, are almost compelled to use adaptive strategies: that is, work intensification and demands for wage concessions. Accordingly, the opportunities for development of training programs linked to workers' learning capacities are likely to be quite limited.

Finally, the garment industry has faced the most difficult economic circumstances of virtually any sector. While conditions vary drastically between the larger men's apparel factories (whose managerial practices and worker learning issues are very much like those of the auto parts factory workers above), the smaller contract shops and the growing numbers of home-workers, overall this is a sector that seems to have institutionalized a range of "cut-throat," adaptive strategies while ushering in the rebirth of sweat-shops. As the site chapter points out, the sector was ravaged by liberalized trade policies that saw garment sector employment drop by more than one third in the early 1990's. With the exception of basic and competitively essential technological changes such as CAD/CAM in the men's apparel factories, the political-economic climate has made the innovative managerial strategies seen in some of the other workplaces impossible here – most obviously so in the case of home-workers. The effect on workers' learning is likely to be quite negative, with the influence of low wages and low managerial commitment to training programs combined with the spatial fragmentation and social divisions of the labour force all serving to discourage employment-related worker learning. While factory employment levels may have increased with comparative low-wage access to the U.S. market in the wake of the continental free trade agreement, there is little indication of more innovative training strategies.

These structural and formal program features provide criteria for selection of these industrial sectors and specific employment sites, and for offering corresponding predictions about workers' actual learning practices in relation to their paid employment. But many other factors are potentially related to workers' learning. Differences in age, sex, ethnicity, technical skill competencies and prior educational attainments may all influence the incidence, content or effectiveness of learning activities. At each research site, we have conducted in-depth interviews with key union informants and with random selections of workers with all of these background features. The findings discussed in the following chapters are intended to be exploratory and illustrative rather than representative of workers generally. But, at the very least, the common patterns and some of the differences suggested here should both provoke further serious research on workers' own learning activities and clearly demonstrate that much of workers' knowledge, skill and continuing learning practices do, in fact, represent hidden dimensions of the knowledge society.

Chapter Outlines

The remainder of the book is organized in three parts. In Part I, we contextualize our method and conceptual approach to studying workers' knowledge and learning practices. Chapter 1 documents how we went about our research by describing what we call the "hard way": a type of participatory action research intended to fit the traditions, rhythms and distinct cultures of workers and their unions. We detail our experiences and provide a running, reflexive critique. By the end of this chapter we arrive at a critical understanding of the importance of alignment with and commitment to the standpoint of workers. Next we suggest an explanatory perspective on learning that is suited to understanding the practices of subordinate groups such as the working-class. Our critique of "deficit theories" of working-class culture and learning capacity is elaborated, followed by a critical review of existing adult learning theories and the presentation of a cultural historical activity theory approach which orients much of our analysis.

The case studies in Part II are the core of the book. These five chapters all originated as site reports on our interviews and observations at the respective sites. After the reports were given back to the members in each union local, they were further developed based on feedback from the local. The different thematic foci of these chapters reflect the central concerns raised by the respondents at each site.

In Part III, we offer some comparisons between these groups of workers. Chapter 8 presents an exploration of household and community learning, practices that are rarely if ever discussed in research literature on learning. This chapter sets the groundwork for further research in hidden dimensions of the knowledge society, and sums up our experiences with interviewees in the process. In the final chapter, we summarize and compare key findings across all chapters and provide a set of recommendations for progressive social research and social action by groups interested in fully recognizing and finding relevant ways to apply the knowledge of working people to build a better workplace and a better world in the new millennium.

Notes

1. See, for example, *Information Highway Advisory Council*, 1995 (vii, 57); *Advisory Committee on the Changing Workplace*, 1997 (5–6); *Speech from the Throne to Open the Second Session of the Thirty-Sixth Parliament of Canada*, 2000 (1–4).

2. For conceptualization of the multiple dimensions of underemployment and a review of mainly survey-based studies of education-job requirement matching in Canada and the United States, see Livingstone (1999).

3. For a fuller discussion of definitions of informal learning, recent general studies and continuing research challenges, see Livingstone (2001).

PART I

RESEARCHING LEARNING AND WORK

CHAPTER 1

Starting with Workers and Researching the "Hard Way"

co-authored with D'Arcy Martin

Introduction

> When I was first approached, I have to say honestly, I thought the idea to have more information was good in principle, but when you try to actually do something it's harder to get going. When I talked to you and we sorted out who we might be interested in interviewing I was still sceptical, but when I talked to the guys they seemed willing to give it all a try – at least some of them. After the thing took off though, and people got to see that what they were saying was going somewhere then things got even better and I think the final presentation and the booklet and everything was a cracker.... Others became much more interested, and here we are. (Union Local President)

This chapter explores the social dynamics, research design and methodological issues associated with our investigation of the hidden dimensions of the knowledge society and working-class learning. The discussion pivots on a form of progressive engagement with the labour movement. We draw on a reflexive and self-critical analysis of how we proceeded and the methodological choices we made.

In the most general terms, our experience has taught us that the most effective *and* progressive means of undertaking social research with workers contrasted with traditional academic modes of research. In fact, our methodological beliefs are closely intertwined with a commitment to an understanding of human agency in the world. This resulted in our stance that if people could be and were active agents in the social world, so should they be in the process of representing that world. This stance, it should be noted, is indebted to the tradition of Action Research (e.g., Fals-

Borda, 1991; Fals-Borda and Rahman, 1991 Reason and Bradbury, 2001), though we wish to add some additional points of clarification suited to researching with the labour movement specifically building on the progressive work of others who have expressed similar orientations.

We return to this theme again in our concluding chapter, but to our mind, this type of commitment places us firmly in a particular "camp" of work-based learning researchers; that is, among those who engage directly with workers and their organizations as both a methodological and epistemological point of departure (e.g., Foley, 1998; Worthen, 2001; Gereluk, 2001; London and Wilson, 1990; Simon, 1990; Spencer, 2002; University and College Labor Education Association, 1977). The remaining two camps, in our view, are represented by, on the one hand, various forms of independent research approaches in which workers practice is taken seriously, but within which working-class standpoints are not specifically articulated (e.g., Billett, 2001; Engeström, 2000; Boud and Garrick, 1999; Darrah, 1996). And on the other hand, the camp in which the standpoint of human resources, management and capital are openly articulated as the point of departure (e.g., Betcherman, Leckie and McMullen, 1997; Marsick and Watkins, 1990; Senge, 1990). In this sense, our methods are closely intertwined with a conviction that learning should enhance working people's individual and collective agency in the social world and also in the process of representing that world.

The "Working-Class Learning Strategies in Transition" (WCLS) project was aimed at assessing the full range of learning practices that union members and their families undertook during economic restructuring in Canada throughout the 1990s. We used in-depth interviewing and selective ethnographies as well as basic document analyses as our methods. Across the five research sites we interviewed over 120 workers (see Appendix which provides summary profiles for all respondents) conducted a variety of focus group sessions and engaged in countless information-sharing, feedback and informal discussions to better understand the patterns that were, for both researcher and researched, emerging. Moreover, we inserted the knowledge of these emerging patterns, in many cases as they were emerging, into the activities of workers in the union locals themselves to assist collective bargaining and policy development, and to aid in mobilization campaigns, and so on.

Interviewees were recruited through their union local. In cooperation with union representatives we developed purposive samples that helped assure coverage of key social variables. These social variables included: gender, race/ethnicity, educational attainment, union activism, age, seniority and, of course, occupation. After establishing research terms with the union representatives and union sponsors (discussed in more detail below), researchers typically made presentations to the rank-and-file membership of the union local at a monthly meeting, followed by individual contacts with potential participants through their shop steward, and finally meetings with interviewers. In some cases, a language barrier was partially overcome through our recruitment of special interviewers who could converse in the workers' first language. This helped us begin to grasp the experiences of workers from specific ethnic standpoints not otherwise represented by our core research team (who each spoke English as a first language).

We conducted two distinct sets of interviews with each worker. The first interview was a focused, semi-structured discussion of learning and work practices that took place over the previous year. The second interview was a broader "learning life history" interview that made space for people to put their current practice into the context of their broader life span. Together, these sessions lasted between two and eight hours per interviewee. This allowed both researcher and researched to contextualize and understand the broader, as well as the more immediate, dynamics of their work and learning. The ethnographic contributions were based on field notes from the interview meetings (in the home, the workplace, neighbourhood coffee shops or bars, etc.) as well as notes from our many different visits to plants, union halls, etc. Close partnerships with unions were an enormous aid in gathering relevant documents, including collective agreements, employer announcements and correspondence with the union, as well as research previously carried out by the union or its allies.

Our approach to the phenomenon of "learning" is based on a socio-cultural model, but in which we wish to highlight the relationship between power and knowledge. A careful recognition of social standpoints is central to this, as is the recognition that the patterns of distribution of all sorts of resources (cultural as well as material), that is power, affect knowledge production practice *and* people's account of those practices within interviews. Some inter-class comparisons (between manufacturing workers and corporate executives) in Sawchuk (2003a) provide substance to this claim, as do the debates that have surrounded the emergence of participatory action research over the past three decades (see Reason and Bradbury, 2001). People's accounts of "informal learning" in particular are susceptible to the effects of power. In short, informal learning is recognized most readily (by interviewers and interviewees alike) when these practices are based in a social community that legitimizes them. Dominant class groups find this legitimacy all around them, but for subordinate groups it is more difficult to come by and must be actively created. Throughout this book, and particularly in the closing chapter where we compare patterns of learning across the different research sites, there is a "double" power-relation that shapes our analysis. The analysis revolves around the stability of working-class communities: stable because of their ability to constitute an alternative language or discourse within which a subordinate standpoint has a (positive) place, and stable because of the accumulation (and collectivization) of scarce material resources. Hence, interviews with auto workers allow an account of an expansive array of learning practices, because of their actual practice *and* their ability as members of a well-developed cultural community to legitimize and report these practices.

These and other related issues must be taken into account in undertaking research of this kind, and this gave rise to our notion of doing research the "hard way". The hard way describes the deeply personal and politically engaged work that social research can become when commitment, care and perseverance are present. These ideas emerged from our discussions with each other as well as our discussions with members of the labour movement with whom we carried out our research. In these discussions, we spoke about the key events and our developing approach to research of this kind. As part of the research team's effort to think critically about our own practice, we went so far as to tape-record, transcribe and analyze our discussions, and some of the lessons we learned appear in the form of stories shaded and set off from the main body of the chapter. These stories fuelled reflection.

Research Methods

"The Hard Way"

Our research findings and method affirmed the fact that working people are agents of their own activity, rather than "human resources" in someone else's economic machine. Though they are constantly faced with challenges of this "machine," in site after site we found people sharply critical of existing power relations, reflective on past efforts to change those relations, and aware of the learning required to shift the processes that have worked against their interests.

Researching with Diverse Workers

I've got a long history with the local in which I interviewed, which goes beyond this project. My links are directly with the business agent of the local. The outstanding issue for her was relevance. In which way could this research project be made integral to the work, to the very survival issues, that the union itself was facing? I don't think that we were actually able to respond adequately to that, because of the way the research itself

> was designed. That calls for a little reflexive thinking on our part, about how we integrate the research design process itself. Specifically, we positioned garment as a sector in decline, a point of contrast and difference vis-à-vis the other sites. In terms of interviews themselves, I didn't do family interviews directly. It was difficult to get an interviewer who could do the interviews in the first language of the family members being interviewed, which created a set of problems on their own. I'm sounding a bit negative here, critical I guess. It was hard in the course of the family interview to move beyond the positioning of the individual who was being interviewed. (Union Contact/Researcher)

These findings work against dominant images of workers as passive, complacent or hopeless, in part because of the theoretical conceptions we brought to the study and in part because of the relations of mutual respect that had already been developed before data collection began. They also reflect, however, the learning that we did as "researchers" and "researched" along the way. It is often said that students teach teachers. In our process, the interviewees often challenged and surprised us, generating new research approaches and deepening our understanding of the relations between the academy and other workplaces.

Much of the dominant literature in one of the fields to which our project was related (adult education) builds on research with university students. Stephen Brookfield, a prominent writer in the field, explains that this is understandable for a number of reasons, many of which we outline in this book. However, allowing a certain population (i.e., students) to stand in for the average adult learning skews perceptions of the phenomena of interest if sweeping or general comments are made. Some of the same reasons that make university students easy to work with can be found when researchers go through a large employer to gather a sample on which they can base sweeping comments about the working class. Since our project involved employed workers as the key population, and chose to address them through their own union structures, we found out to what degree we had chosen the "hard way" of doing research.

Doolittle, PhD

This was the first research project I have ever been involved in, so this relates as much to my own learning as anything else, my own education as a researcher. I was a fresh student when I started on this project on working-class learning strategies, very new to academia. I remember at one of our early meetings I asked if I could interview someone I knew. I had identified a friend whom I knew from my days in that same workplace. The answer naturally was yes, no problem, but this wasn't research as I knew it. To me that smacked of bias, loss of objectivity, which ran counter to the grain of everything I had learned as an undergraduate. But the answer was such a matter of fact yes, I was frankly surprised. The interview itself was difficult. It was cut off at the insistence of the worker. I had given him a copy of the interview schedule ahead of time, to build trust, to say "you'll have a copy and I'll have a copy, there will be no tricks." I remember once at a research meeting we discussed whether people were sticking strictly to the interview schedule, and I was the only one who said yes. Anyway, this person looked over the interview schedule, then tossed it away, saying "this is all Doctor Doolittle bullshit." (Researcher)

We can explore this notion of the hard way by examining its different dimensions. The first dimension relates to availability. University students are relatively time-rich and cash-poor. For this reason, they can be accessed physically and, if the research budget allows for a fee, they can be fairly easily motivated to participate. Similarly, if an employer grants access for researchers during working time, some of the same dynamics apply. But our conversations were mostly outside paid time and often outside the workplace.

The second dimension of the hard way relates to issues of homogeneity. That is not to say that all university students share the same sets of social identities, but certainly their class and racial composition is increasingly consistent as cutbacks drive up tuition fees and political leaders retreat from discussions of equity and accessibility. Generalizations are therefore relatively easy to develop from interviews or questionnaires targeting this particular population. When we turn to workers, in workplaces that are mostly white, male and unionized, some of the same dynamics apply. However, we chose our populations explicitly to cover a broad range of working-class experiences within society. We dealt with workplaces in which class, as well as gendered and racial/ethnic, standpoints had to be addressed. This required us to reorient research tools to fit differing situations, and left us with major difficulties in making generalizations at the end.

A third dimension is familiarity. Here we consider the cultural assimilation of students into university definitions of "knowledge". To most students, the idea that academic research is valuable is taken as a given. Further, the social and organizational networks are such that any misuse of results can be challenged directly. With employed workers whose participation is sponsored by the employer, the argument is at least made that this can be of value to the enterprise, whose spokespeople are engaged in the project. While this does not guarantee honest answers, it does imply that participants will not be puzzled by the very idea of academic research. In our project, workers met us through their union, although not all interviewees were activists or even union supporters. This provided two sets of hoops to jump through: one the experience of members with their union representatives and the other experience of union representatives with academic research.

A fourth dimension was the notion of a "cooperative" attitude. The social construction of academics as "authorities" reinforces a tendency to collaborate and even to submit when students feel doubtful about the particular research enterprise (cf. the studies of subject compliance in administering "lethal" electric shocks to "research subjects" in Milgram, 1974).

Trust and Solidarity

I'm thinking about the difference in discussion that comes out in the interviews depending on how the participant views me. I'll give an example. I think about the family interview I did with a local president. I identified myself as a person newly active within another local of their union, and then we had a completely different relationship. We were speaking in a spirit of solidarity. That really showed up in the family interview, where I got a lot of stories that were a lot more personal, and there was a lot more probing. The kids were more comfortable with me, since I interviewed them too, and actually wound up cutting their hair at the end of it. I was welcomed into their lives. (Researcher)

We need not consider students submissive to realize that a powerful pressure to take the research design as given and beyond challenge exists in the research relationship. Technical arguments regarding various guards to validity or replicability can be used to limit questioning by subjects, with the encounter reinforced by the social and educational standing of the lead researcher. Similarly, when a worker is assigned to cooperate, or encouraged to volunteer in a project by an employer, a layer of "soft" coercion necessarily underlies the surface of consent. In our study, there were few tangible benefits to workers for participating beyond the opportunity to reflect on and better understand their own practice. Indeed, participants often had to sacrifice their time and energy to be involved. This is particularly important when asking people to express the ways that they construct their lives, to recount their self-activity, to draw upon implicit or hidden dimensions of their home and work lives, and to apply a lens of learning when making sense of experience.

A fifth dimension involved the idea of "efficiency". For university students, the role of respondent to a questionnaire or interview is acceptable, the framework, the social situation and phrasing of questions is recognizable, and the need to consult with others before responding to issues raised is minimal. This makes student subjects comfortable and expressive within the research process. Similarly, when a worker is supported by an employer to participate in a study, the assumptions embedded in the research instrument will already have been tested in terms of their consistency with corporate goals. Subjects, then, represent only themselves and are in this sense "free" to speak. For our process, people often wanted to refer back to colleagues to validate their opinions, and repeatedly voiced suggestions of others whose opinions might be different or somehow more valuable than their own. As a reflection of our own conceptualization of learning as a social act, rather than an individual one, workers tended to respond from the standpoint of a group, union or community member.

In these five ways, then, our research process was likely to generate conclusions different from the main body of literature on adult learning and workers' lives. Put bluntly, we found it highly inconvenient to pull back from dominant sets of research practices to engage in ethnographic, situated, intensive and culturally respectful interview research with workers. We paid a price, in the painful slowness of our progress and the difficulty of gaining sufficient diversity and comprehensiveness to feel confident in the conclusions at which we arrived. Nonetheless, the effort invested in doing research the hard way also produced its own types of benefits.

The Reluctant Subject

This fellow had very little interest in doing the interview at all. He was a kind of quiet and sullen guy. Amazingly enough, he came to all the monthly meetings at the local which I attended. He never participated, but he sat in the back. Finally someone suggested as a way of bringing him into the local why didn't he get involved in this interview. He finally agreed after a little bit of grumbling. So I was kind of uncomfortable, not knowing what would happen in the interview itself. He gave an awful interview. It was brutal. He didn't want to engage with the questions at all, but he showed up. It was a very strange situation. He was very concerned about where this research would go, who was doing it, who in the local was behind it, what it aimed to achieve and so on. At that point, we just abandoned the interview schedule, and went on to discuss these very issues, of who would use the research and for what purposes. It turned out to be a very important political move inside the local. He was a guy who was greatly respected in the workplace. He was the kind of person who didn't jump into things excitedly, which I could obviously see. What was important was that the interview was really helpful for getting initiatives going based on the research, which I think is pretty rare in academic life. I ended up giving him a ride to the subway station after that, and word spread through him. I didn't know it at the time, but that was the seal that said something was actually going to come of all this work. (Researcher)

Facing the Challenges of Researching the "Hard Way"

In designing this research, we consciously chose to deal with a wide range of workers, in settings comfortable for them, and emphasizing their collective affiliations, especially, though not exclusively, with their union. From the beginning, we expected that this would require patience and flexibility. We were right.

Our subjects were not easy to access. Due to shift work, the double day of paid and unpaid work, and for many, due to the need for a second job, our subjects were "time poor"; and among them few were as "time poor" as the union activists who also volunteered time to represent their co-workers. Nowhere were these factors more evident than in our activities in the garment and small parts manufacturing sites where time and human energy were particularly precious and scarce. This resulted in meetings often being cancelled or postponed. It required in some cases that specialized interviewees be recruited, but more generally it required that researchers be empathetic and fluid in scheduling, and open to "interruptions" by co-workers, family members, and even waiters and waitresses depending on the location of the conversations. Here the key contacts in each research site were particularly important, encouraging people to remain involved, to reschedule rather than drop interviews, to encourage the researchers in doing the same. The stubborn insistence on accessing busy people was because the research team was clear that the classically "self-directed" and highly motivated learners in our sites were a relative minority. Union activists and what unionists call self-motivated, "bootstrappers" play a significant role in the life of these workplaces, but limiting our interviews to such people would certainly limit any conclusions we might reach.

Our respondents were heterogeneous. Life is very different for workers in an auto assembly plant compared to a clerk in a college admissions department or a home-worker in the garment industry. This makes it difficult to find significant generalizations, or to use identical research tools.

A Network of Trust

What strikes me is the importance of social networks in what we have done here. We are actually in some kind of loosely configured community of left-wing intellectuals that think unions matter. Then we disagree on a hundred other matters, but we are personally connected, so it affects how we show up at meetings. Today I may meet one of the team for the first time, but other people I have worked with in a whole variety of situations – writing stuff together, thesis research, course participation – we have had this range of relationships, so that what we do now actually impacts an existing relationship. What we do in the relationship also impacts the research. So it's much more of a venture within a community, an intentional community of some kind. That distinguishes our research process. So a right-wing and positivist researcher would have a hard time participating in this project. That's not an ideological statement, it's a statement of what it has taken to pull this research together. The fact that I respected other people's past work had a lot to do with wanting to participate in this project in the first place. I'm not sure I would have participated at all if approached by someone else from a different department, and I sure as hell wouldn't have turned over a direct link to a local union in the way I did. So I see issues of trust, of community, and how intellectual work plays out there. (Union Contact)

Equally challenging was the fact that our interviewees were suspicious of us. The ongoing experience of working-class people in being ignored, patronized, misunderstood and misrepresented by the dominant voices in our society made them cautious about opening up in the face of the forms of "cultural capital" carried by academic researchers. A skepticism about "book learning," akin to that described by Paul Willis (1977/1979) in his classic research on working-class boys, was widespread in these sites and also weighted into their unease in the situation. And in our research, given that this was union-sponsored research, there was no incentive from the employer for them to participate. Underlying this reflex was also a sense that, should personal confidentiality be broken in any way, the chances of holding anyone accountable for how information was used would be tenuous. Again, although participants were aware of our guarantees of anonymity and confidentiality, there was palpable caution in trusting academics. Overall, the link was from the worker to the union local, then to the key contact person or sponsor, then to the research coordinator, researchers and the university. For rank-and-file workers, this path appeared scarcely direct or sturdy. To address this, careful choice of the key contacts was needed. These were people who "bridged" the two worlds, with a link to the academy and the workplace and union. They had to be trusted and respected on both sides. The choice to work the "union network" rather than the "management network" was central to the design of the study. The labour movement, in the words of former Canadian Labour Congress president Dennis McDermott, is a "network of trust." While it has formal structures of accountability, initiatives are often launched informally and draw on past common efforts. In the absence of personal trust, unionized workers are quite literally expert in passively sabotaging new initiatives, and hence the choice to work with people who were trusted by the lead researchers *and* by the union leaders in the site was fundamental to the success of the project.

Linked with the caution and suspicion of workers were forms of rebellion and resistance. In fact, the more active people were in their union, the less they conformed to the image of North American workers as increasingly complacent. Rather, they were resistant to authority figures, including academic ones. Our project chose to treat resistance as being full of information and energy, as a force to be developed rather than ignored or undermined (as is often the case in Human Resources–based research as far back as the original Hawthorne Works research [cf. Martinez and Eston, 1992; Schied, Carter, Preston and Howell, 1997]. In the course of our interviews, rather than brushing aside remarks about power relations, for example, these issues were probed and expanded. Beneath this stance is a political choice to work supportively with workers who see their interests as in some way clashing with those of their employer, with the employer's academic allies, and with the dominant classes in society as a whole. This does not mean putting words in the mouths of interviewees, but rather choosing to record and explore the statements that, in other projects, might go unregistered, ignored or erased.

Surviving Union Politics

The story I'll tell has to do with thinking about keeping the research alive over a longer period of time. It recognizes the importance of having multiple contacts, who understand and believe in the project. The current executive can change, for a variety of reasons, and you may be up the creek without a paddle, in terms of being able to continue to engage with the local. Even if you've done useful work previously, that some people have found valuable, unless you have a number of people who are prepared to support your work on an ongoing basis, speaking in the local forums, your work may come a cropper, and it may not have any sustainability in the local. Just a footnote on that. Precisely if it is seen as useful work, a positive accomplishment of a regime that is no longer in favour within the local, it can be junked, not because it isn't good research but precisely because it is good research. So the better your work, the more politically vulnerable you become if things polarize. (Researcher)

The final challenge of the hard way was rooted in the collectively situated nature of unionized workers. This means that people had direct experiences of operating in a unionized environment that demonstrated the power and necessity of collective perspectives and at the same time the negative side of managerial practices that sought to "divide and conquer." For those active in their union, this means a caution to express divergent views, a desire to consult with others or a willingness to see their personal experiences in the light of that of others. Recognizing too that unions are a political organization with internal rivalries, it becomes important for the researchers to try to involve different, sometimes competing, union factions. After all, depending on the results of the research there may be credit or blame to apportion with the union local. Our choice was to strengthen solidarities, however fragile, rather than undermine them. Our commitment was to situate internal clashes within the wider social and political mission of the labour movement, irrespective of what the next union election might hold.

Democratic Knowledge Production: The Importance of Social Standpoint

The old question, Which side are you on?, cannot be avoided. Openly taking sides at the start, when it is necessary to be on one side or the other, is not only the radical thing to do. It is the honest thing to do. And who, but those whose interests are hurt by the truth, would want us to be dishonest? (Ollman, 1993: 177).

With the notable exceptions mentioned above, there is a relative lack of scholarship that looks at how working-class people and their unions currently organize and carry out their own learning, and the possible role, if any, for academic "fellow travellers" in this development. We want to focus now on the role of the union local in the development of educational research with an emphasis on the following: the organizational positioning of educational research within organized labour structures past and present (noting also that "education" and "research" are usually separate entities within unions); the connection between unions and academics; and finally, some general lessons to be taken from these reflections. By addressing these factors, we will see how the position of educational research within the organizational structure of the labour movement tends to prevent the full development and use of research by workers and for workers. Of course, as discussed earlier, this organizational positioning serves an important defensive function that protects workers and their unions from exploitation and helps maintain control over information and the development of knowledge. However we will also see how shifts in research practice involving the type of "Union Sponsorship" as described above can contribute to a different research dynamic that benefits the workers at the local level. We mean to speak directly to the potential of research processes and collaboration between academic researchers, workers and their organizations, and how this specific intersection informs not only the action research tradition but theories of knowledge and method as well.

Research is important to the survival of the labour movement. By this we mean that it has been understood for some time that creative adaptation to change is essential to the maintenance of union density and the health of the labour movement generally (Chaison and Rose, 1991; Kane and Marden, 1988; Kumar, 1993). But while research is important to the labour movement, good sense tells us that the way research is done, the questions it asks and the interests it serves makes some research more relevant to workers' lives than others. While we can dicker over methods and the virtues of "professional" versus lay analyses, what we really cannot argue over is the fact that research is *always* undertaken from a particular perspective or standpoint. Standpoint analysis is important because it reveals the operation of particular frames of reference and illuminates the mechanisms by which we apply (often presumed) systems of relevance. When researchers do not recognize different social standpoints, research often meets the needs and matches the worldview of the researchers rather than the researched. As Michael Parenti has commented, "[s]cience is something more (and less) than the dispassionate pursuit of knowledge. How scientific information is shaped is often predetermined by the prevailing ideological climate." (1996:221) The worst research remains painfully ignorant to these issues, and imagines itself as having access to a type of objectivity that is, in fact, impossible. There are valid forms of objectivity, but this traditional, positivist variation that lingers in the social sciences is something that Donna Haraway (1991) has referred to as the "God Trick:" a view of anything and everything from nowhere and everywhere. Defenders of the traditional "disinterested observer" typically refer to research that openly proclaims its social standpoint, interests and imperatives as "partisan." This provokes a question for most readers: How exactly are issues of standpoint and research bias related? Sociologist Howard Becker spoke about it this way,

> When do we accuse ourselves and our fellow sociologists of bias? I think an inspection of representative instances would show that the accusation arises, in one important class of cases, when the research gives credence, in any serious way, to the perspective of the subordinate group in some hierarchical relationship ... We provoke the suspicion that we are biased in favour of ... subordinate parties when[ever] we tell the story from their point of view ... when[ever] we assume, for the purposes of our research, that subordinates have as much right to be heard as

superordinate, that they are as likely to be telling the truth as they see it as superordinate, that what they say about the institution has a right to be investigated and have its truth or falsity established, even though responsible officials assure us that it is unnecessary because the charges are false. [In other words] we provoke the charge of bias, in ourselves and others, by refusing to give credence and deference to an established status order, in which knowledge of truth and the right to be heard are not equally distributed. (Becker, 1970: 125–27)

To clarify, none of this has anything to do with simply "telling people what they want to hear," or the type of relativism that dislodges any hope of developing a shared understanding of the world, both of which are common charges of this type of approach. Rather, it deals with something more simple and more important: the recognition of one's own position in a social system (as an academic but also as a classed, gendered and racialized person). A critical approach to social research thus depends on genuine attempts at enunciating the perspectives of researchers and researched, the contexts of research, the funding, projected uses of the information, and other issues that form the foundation of research but which are typically thought to be somehow separable from it. This is a critical approach that emphasizes the importance of workers (and researchers) keeping their "bullshit meters running" (Sawchuk, 2000: 202). Intensive local union member participation as a method of producing research from a union/working-class standpoint can therefore be an important component for the production of relevant and quality research for labour, and this is the approach we used with this research. In this sense, what is needed is not more research *on* working people, but rather more research *with* and *by* working people on issues defined from their own, diverse standpoints in the world.

Research from subordinate standpoints can contribute to the developed informal and formal networks of knowledge production that workers already have. It is this type of research, as Haddad (1993) suggests in her comparative research on labour unions across Canada and the United States, that correlates so closely with the stability of particular national labour movements. It indicates their ability to adapt and provide their own creative responses and alternatives to the initiatives of capital.

In regards to the linkages between working people, unions and universities, however, the most basic lesson learned is that creating and sustaining real participatory collaboration is not easy. From the beginning of our research, we organized towards a series of "give back" sessions that aimed to have the work tie directly into the activities of the locals. This could be tricky, as we've noted, particularly when intense local politics wash across the sessions. Others considering similar dynamics and concerns have voiced additional difficulties. Forrester and Thorne (1993), for example, comment on research links in relation to a conference designed to establish and celebrate union/university research collaborations. In a book documenting the conference they comment that "[o]n the whole, however, the significance of the contribution by trade unionists to the Conference is not reflected in the contents of the book: most of the contributions are from professional researchers. This is, in itself, an indication of some of the difficulties inherent in the relationship between trade unions and researchers." (ix). Later they quote a rank-and-file union member participating in the conference.

> Here was a conference aimed at developing links between the worlds of professional research ... and the world of trade unions. Yet it is set up in a way that will alienate not only trade unions but also the entire population of non-professional researchers. (x)

The point here is that while the traditional academic mode of knowledge production (e.g., scholarly papers in scholarly journals, scholarly presentations, and scholarly conferences) may have its uses, it is probably not sufficient for supporting the development of union/academic research linkages, particularly at the level of the union local and rank-and-file membership.[1]

As we indicated from the beginning, there is no way to complete this chapter without mention of the field of Action Research. This is a research approach in which, in its most basic form, study is undertaken *within* and *contributing to* activity undertaken in solidarity between the researchers/ researched. In addition, the data collection itself (in-depth, life-history interviewing) had an enormous effect on building the commitment of all concerned within the research process as a whole. As several local officers commented, these methods put a "human face" on the research, attached "practical" outcomes to the research, and, in general, formed a positive basis for the ongoing work at hand, as well as future research. Again quoting from Forrester and Thorne (1993),

> The agenda is less likely to drift away from their concerns and the results have both the legitimacy and relevance accorded by lay member involvement. Implicit in the research activity is a restructured relationship between researchers and workers to create teams capable of combining the research skills of academics with the knowledge of lay activist union members ... Research can only become part of a programme of genuine empowerment when the research process itself is democratized. (4)

Concluding Remarks

One of the underlying arguments of this chapter is that not just any sort of research can produce the positive effects workers and progressive researchers want most. We claim that projects interested in the lives of subordinate groups are most successful when they operate from the standpoint of and/or in solidarity with these groups. In the first half of the chapter we discussed the intricacies of how, in practical terms, this can be accomplished and the challenges that this type of research must face. Central to these intricacies was the role of Union Sponsors. These sponsors are central, in this and many other forms of social research with unions, to the establishment of dialogue and negotiation between outside researchers and a union local.

However, the concluding thought with which we would like to leave you involves thinking about the relationship between research and social action as a dialectic and, more specifically, as a form of praxis. In other words, knowledge and action must be linked; that is, knowledge building can be done in the course of contributing to forms of social action. In this case, the value of research is realized in both its practice *and* in its production/dissemination beyond the local. However, what we are really talking about is beyond any purely instrumentalist exchange between unionists and academics. It is perhaps best described in terms that Raymond Williams has referred to as "alignment" and "commitment" (1977: 199–205). Effective social research with the labour movement, to our minds, means moving beyond instrumental motives to arrive at a deeper understanding of our shared standpoints as wage-earners, and possibly as people who see potential for things to be better. This entails moving beyond commitment to one another to see the ways in which we may be deeply aligned. In other words, it means coming to understand how the interests of workers in different economic sectors show important overlap with academics working in the higher education system.

Academic rhythms, workplace and union rhythms do not, and did not in our project, run in a synchronous pace. Our choice was to privilege the latter over the former in most cases, and this should be considered as part of the overall research design we advocate. In general terms, this willingness to adapt was based on the recognition that good things take time. Building trust is a lengthy process. For those seeking to carry out research of the "hit-and-run" variety, simply put, the labour movement is no place to turn. Moreover, for those choosing to balance their training as traditional intellectuals with political commitment and alignment with the labour movement and its members, the process takes years to deepen and ripen. In this sense, we suggest that research be viewed as one moment in a long-term process of helping to release the eloquence of workers, recording and amplifying their words and experience. Our goal was to bring to the specific academic literatures a validation of the experience of workers as active agents of their world, as individuals with strong collective supports, as curious and lively and diverse voices in a time when the corporate monotone increasingly threatens to drown the field.

Notes

1. Related to these tensions, our own labour-positive conferences developed this set of conventions for maximizing the effectiveness of union participation: a) labour representatives on planning committees; b) flexible presentation formats such as roundtables as well as physical settings that supported both the time constraints of unionists and the oral culture of trade unionism; c) establishing caucus sessions each day for union participants to share experiences and perceptions of the conference; d) allowing free access for union researchers who typically find it difficult to justify the financial expenditure (beyond the time off) to contribute; and e) inclusion of at least one labour representative on any closing, summary panels for the conference.

CHAPTER 2

Beyond Cultural Capital Theories: Hidden Dimensions of Working-Class Learning

[A]ll policy deliberations that do not draw out fully the range of possibilities which are available for action and which science can aid in revealing, can be considered as guilty of non-assistance to oppressed people. Pierre Bourdieu and Contributors (1993: 145 [authors' translation])

The Indian peoples of Oaxaca in southern Mexico, to take one example, have flourished, as have their places, because of their traditions of teaching and learning. Their diverse cultures have continued to be enriched despite the abuses and interventions they have suffered from all kinds of Outsiders.... [They] have kept regenerating their language and culture, while coexisting with, as well as resisting, their colonizers' universalizable truths. Their evolving modes of cultural coexistence protect their pluriverse, adapting to each new condition of oppression and domination without losing their historical continuity.... When studied from up close, we discover with others that those who become addicted to classroom instruction end up losing real opportunities for gaining the knowledge and skills with which communities endure and flourish. Prakesh and Esteva (1998: 3, 4, 7)

Introduction

Through the history of societies in which social hierarchy has influenced the distribution of material resources, ruling groups have generally attempted to characterize the thinking and acting capacities of other social groups in condescending and inferior terms. Those in dominant social positions have developed elaborate ideologies and theories to legitimate their own superiority and the lesser rights and motives of those beneath them.

The labouring classes, women, minority racial and ethnic groups and those designated as disabled have all typically faced both restricted access to empowering material resources and invasive portrayals of their inherent intelligence and other abilities as naturally and inevitably diminished in relation to ruling classes, men, dominant racial groups and the "able-bodied," respectively. But even in slave societies in which the material constraints have been most severe, subordinated groups have generated their own subcultures which resist, reject or invert these dominant ideologies and theories (e.g., Genovese, 1971).

With the generalization of wage labour under industrial capitalism, working people gained sufficient freedom and economic strength in early-nineteenth-century western Europe to sustain independent collective organizations (e.g., trade unions, political parties). The initial emergence of historically and culturally specific theories of society and social identity that recognized group differences as grounded in the context of social relations was coincident with the development of these oppositional organizations and often closely aligned with them, most notably historical materialist (or Marxist) theories and Communist parties. The twentieth century saw the proliferation of widely accessible means of production and dissemination of information and knowledge (e.g., telephones, computers), various challenges by organized labourers to generalized private capitalist ownership as the "natural" form of productive enterprise, and the sustained emergence of feminist and civil rights movements. It also saw the production of more nuanced theories of the active development of social selves in diverse historical and cultural contexts to challenge still dominant notions of a dichotomy between the individual and society, of the human soul and personality as biologically or spiritually innate, and of established institutional orders as remote from individual activities (see Burkitt, 1991).

More generally, the increased freedom of association and communication during the twentieth century in Western societies encouraged the development of diverse public spheres through which the interests and views of previously silent social groups may be articulated (Negt and Kluge, 1993). Increased freedom of expression has not been a linear process throughout the century in all Western societies – witness the emergence of fascist regimes during the Great Depression and the recent aggressive promotion of neo-liberal ideologies of "free market" individualism – but multiple overlapping public spheres animated by collective organizations variously constituted by those in subordinated class, gender and racial positions are now discernible. Postmodernist discourse theories, for example, may be understood as a response to these freer conditions in societal contexts that remain hierarchically organized and oppressive.

Ironically, contemporary efforts by scholars critical of social oppression to document the logic and effects of dominant ideologies have often served to portray their oppressive effects as greater than they really are, because comparable attention to the ideologies and practices of subordinate social groups has not occurred. Cultural reproduction theories such as cultural capital theory (e.g., Bernstein, 1990; Bourdieu, 1984;) are prominent examples that will be examined here in relation to the actual learning practices of organized working-class groups.

The Cultural Capital Bias

Contemporary cultural theory has been developed largely within bourgeois dominated academies in splendid isolation from organized working-class practice.[1] Academic portrayals of working-class–based cultural practices have thus generally missed the creative agency and original features of the class culture, or at best conveyed them in disembodied and fragmented ways.[2] Most recent contributions to critical cultural theory by Marxist and other scholars have generally been produced in academic settings without sustained practical engagement with the organized working-class.[3] Strong social forces, including capitalist control of increasingly pervasive mass media and the disruption of working-class communities, have threatened the working-class capacity for concerted collective representation. But sympathetic academic analysts' continued remoteness from the working-class's own collective agencies has also aided and abetted the undermining of positive working-class self-perceptions through their production of excessively disembodied and fragmented versions of current working-class cultural practices.

Human capital theories generically assume that human knowledge and skills are analogous to any other commodities offered for market exchange. In general, the more individual investment one is prepared to make in acquiring educational credentials, the greater the market reward is likely to be (Becker, 1993). Among the most influential examples in the field of education and learning are the "cultural capital" theories of the reproduction of social inequalities through schooling developed by Pierre Bourdieu and Basil Bernstein.[4] Both these eminent scholars have drawn lightly on the Marxist tradition in developing their conceptions. Bourdieu and Passeron's (1977) notion of "cultural capital" was developed by analogy with Marx's analysis of capital as an accumulation and reproduction process, while Bernstein (1990) has noted the comparability of his restricted and elaborated language codes to the work of the Soviet psycholo-

47

gist, Lev Vygotsky. While a great deal of contemporary learning theory takes a class-blind and individualist perspective, both these social analysts have developed models of class differences grounded in cultural sensibilities and linked them to differential social effects of schooling processes. A critical appreciation of their contributions is needed in order to move beyond the limits of cultural capital theory. In both Bourdieu and Bernstein, the primary emphasis is placed on the general cultural knowledge, sophisticated vocabularies and precise information about how schools work that children from higher status origins acquire from their families. The possession of these cultural tools leads to these children having greater success in school relations than do working-class children. Such cultural theories offer considerable insights into the discriminatory schooling conditions faced by working-class people. In Bourdieu's case, as in human capital theories, human learning capacities are equated with capital assets. Children of the affluent classes, who have acquired familiarity with bourgeois cultural forms at home (through exposure to their parents' knowledge and manners, as well as linguistic forms) are seen to possess the means of appropriating similarly oriented school knowledge relatively easily. Working-class children, in contrast, find their unfamiliarity with these cultural forms to be a major obstacle to successful school performance. Bernstein makes similar arguments primarily in terms of language codes, with upper-middle-class children seen as possessing more elaborated codes for abstracting and generalizing from school curricular materials. Both scholars, with the aid of teams of colleagues, have done extensive empirical verification and refinement of their models, deepening both their complexity and their insights into the discriminatory cultural processes that operate against the working-class in many schools. Both theorists expose the dominance side of cultural reproduction in excruciating detail. At least in this limited sense, their contributions are comparable to those of feminist and anti-racist scholars who have critically exposed the dominant codes and structures of patriarchal and racist cultural forms (e.g., Spender 1980; Said 1993).

However, Bourdieu and Bernstein were preoccupied with delineating the cultural reproduction of inequality within fixed institutional forms. Thus, their accounts remain primarily one-dimensional, functionalist descriptions of the status quo rather than real explanations of it (see Curtis, Livingstone and Smaller 1992: 6–25). Bourdieu's and Bernstein's theories of class cultures have ignored a central rule of sociological investigation promulgated by one of the founding fathers of sociology they both build on, Emile Durkheim: to understand any social fact, we must study it through the full range of its variation. *In sum, they never comprehend the creative cultural practices, independent education and learning activities, or collective cultural agency of the organized working-class.*

As Collins (1993: 134) concludes in one of the most nuanced critiques of Bourdieu's class-specific research on language and education,

> [Bourdieu's] dialectic of subjective action and objective conditioning.... lacks a sense of the primacy of contradiction, however, and we are left with an account of conditioned strategies for action that overrides the conflictual creativity of interaction-based agency.... [T]he discursive always seems deducible from, reducible to, in a word, determined by something else: class conditions, capital composition, habitus, field effects. There is a truth in this determinist argument, but it is one-sided.... We need to allow for ... creative, discursive agency in conditions prestructured, to be sure, but also fissured in unpredictable and dynamic ways.[5]

Working-class people are always presented as reactive and marginalized in these perspectives, even in the more recent, empirically grounded works (Bernstein, 1996; Bourdieu and Contributors, 1993). For example, as Fowler (1997: 11) observes, Bourdieu has exaggerated the cultural dispossession of the masses and excluded any popular art in his category of consecrated culture, constructing a canonical closure which is too complete and which blinded him to the existence of authorship within these popular art-forms. Similarly, Pearl (1997: 143) argues that Bourdieu's central concept of "habitus," which refers to humans' situated cognitive functioning and sense of possibilities, underestimated shifts in expectations and aspirations among different groups. With some notable exceptions (see Bourdieu, 1998; Bourdieu and contributors, 1993), he can be read as providing a highly sophisticated defence of deficit theory (i.e., the limited capacities of lower-class and other subordinated groups) while ostensibly aligning himself on the side of equity and social justice.

One of the most striking facts of working-class knowledge in capitalist societies appears to be the multiple barriers to its recognition and legitimate use in dominant institutional settings. We have alluded to the multiple dimensions of *underemployment* and their disproportionately high incidence among working-class people in the Introduction. Versions of this theme of wasted knowledge and denied recognition of working people's skills and learning capacities are expressed by nearly all of our respondents. As a young woman public sector worker who participated in our study complains, "My friends are all underemployed. People from my generation, twenty to thirty-five, we're all underemployed. People with university degrees, but they're maids or waitresses. Everybody I know. If not degrees, they have incredible intelligence and are good at a lot of things ... Education and the work world need to be re-thought."

It is equally striking that the proponents of human capital theories of education have virtually ignored the problem of underemployment. We suggest that this is primarily because their conceptions of knowledge and learning processes have focused almost exclusively upon aspects of knowledge that are easily expressed in commodity exchange relationships. Most versions of human capital theory rely heavily on an identity with financial capital. The Marxist distinction between exchange value and use value is highly pertinent here. It is only from a working-class standpoint that the use value of much of working people's knowledge that is not counted in school and labour market exchanges becomes visible and valorized. But even if working-class learning is conceived in the extraordinarily narrow terms of the labour market exchange value of school credentials, the underemployment of working people's knowledge in capitalist workplaces should now be seen as an endemic social problem. In any case, in stark contrast to cultural capital theory, researchers' sustained engagement with working-class–based organizations begins to generate an alternative perspective on their cultural sensibilities and learning capacities.

The Rediscovery of Creative Working-Class Culture

The socialization of the forces of knowledge production (especially through the availability of free voluntary forms such as public libraries, trade union schools and now electronic information networks) is a major source of autonomous cultural production by subordinate social groups. The increasing availability to working-class people of such socialized forces of knowledge production represents a continual challenge to private capitalist efforts (via conglomerate ownership of mass media, commodified information packages) to control the social relations of knowledge production. This opposition between socialized forces and privatized relations of knowledge production is the *fundamental contradiction of knowledge development and learning in capitalist societies* (see Livingstone, 1999a).

The extensive capacity of working-class people for knowledge production and cultural creativity was both recognized and encouraged by many working-class leaders until the 1920s. The most extensive example has probably been the *proletkult* movement in the Soviet Union, initiated by Bogdanov after a split with Lenin in 1909. Shortly after the 1919 revolution, *proletkult* involved about half-a-million workers in networks of workshops, studios and clubs focused on the arts, actively creating a new working-class culture while critically reworking the best of past culture.[6] There are numerous other substantial examples of such independent working-class educational organizations early in the 20th century (see Sharp,

Hartwig and O'Leary, 1989) but by the end of World War II, hierarchically organized workers' parties and trade union bureaucracies typically discouraged or ignored rank-and-file creativity.

The post–World War II rediscovery of this creative agency has been initiated by sympathetic intellectuals operating initially outside both established academies and labour organizations, and often engaged in adult working-class education jobs, notably Raymond Williams and E. P. Thompson.[7] There are now vibrant areas of study of popular literature, working-class autobiography, historiography and ethnography making these subordinated cultural voices more widely heard in academia (e.g., Burnett, Vincent and Mayall, 1984; Denning, 1998; Zandy, 2001).

The basic point is that working-class and other subordinated groups continue to exercise their own creative learning capacities both within and outside dominant class forms of knowledge. The knowledges that they produce and reproduce continue to constitute oppositional cultural forms both within the realm of education per se as well as in the broader spheres of individual and collective informal learning (see Livingstone, 2002; Sawchuk, 2003a, 2003b). But the question of the actual character of working-class learning can only be answered through direct engagement with working-class subjects and their full array of learning practices in the context of contemporary capitalist societies. We will first review contemporary theories of adult learning and then present our own theoretical perspective, which builds on the cultural historical activity theory approach to comprehending learning.

Critical Review of Major Theories of Adult Learning

Our goal in reviewing literature on adult learning theories is to search for conceptual tools to assist in comprehending the learning processes of subordinate social groups, especially the diverse range of working-class people. Unfortunately, we find a prevalent set of presumptive tendencies or themes running across many of the major approaches to conceptualizing adult learning which serve effectively to diminish recognition of working-class learning capacities.

The identification of these themes is not particularly new (see also Sawchuk, 2003a). They are in fact defining features of the modernist world view that developed along with industrial capitalism. We borrow them from Marx's own critiques of bourgeois philosophy. However, applied to the field of adult learning theory, they assume a unique relevance to us here. The themes we discuss across many different learning theories include: a) individualist/cognitivist tendencies; b) universalist/ahistorical tendencies; and c) formalized learning/expert-novice tendencies. Together

these themes hamper the potential of most theories of adult learning to recognize meaningfully the learning of subordinate social groups such as the
working-class. Many adult learning theories individualize the learner,
reify "learning" into a cognitive-physical act, and make invisible the relations of collectivity and co-operation upon which subordinate groups have
historically depended. Furthermore, theories of adult learning that
universalize the standpoint of the learner hide the contradictory, historical
and political-economic relations that constitute working-class life. Finally, learning that takes place beyond the classroom may be of particular
pertinence to subordinate groups whose capacities are often discounted
within these settings; theories of learning premised on conventional notions of instructor-led pedagogy, or narrowly on the relations between expert and novice, do not provide an adequate basis for understanding working-class learning as a complex whole.

Others have drawn together more general reviews of adult learning
theory (e.g., Foley, 1999; Hart, 1992; Merriam, 1993; Mezirow, 1991;
1996; Poonwasie and Poonwasie, 2001; Selman and Dampier, 1991;
Welton, 1995). However, the goal of this section is to select from the
range of theories of adult learning those that potentially have something
to contribute to an analysis of working-class learning specifically, and to
test them for their capacity in these terms. For our purposes, theories of
"andragogy" (Knowles, 1970, 1975, 1977), "self-directed learning"
(Tough, 1967, 1979), "transformative learning" or "perspective transformation" (Mezirow, 1991, 1994), "critical pedagogy" (Giroux, 1983;
Livingstone, 1987), and finally the notion of "conscientization" (Freire,
1970, 1996) all have something insightful to contribute. Nevertheless,
each, to varying degrees and in unique ways, shows links to the dominant
themes outlined above. We conclude this section with a brief statement on
several recent authors who have been more effective in their analysis of
learning among subordinate groups in order to situate our preference for a
cultural historical activity theory of learning.

Andragogy, that is, "the science and art of helping adults to learn"
(Knowles, 1970: 43) has been a touchstone for mainstream, North American adult educators for several decades. For our purposes, its most important original contribution was in its emphasis that adult learning takes
place within a developmental and social context fundamentally different
from that of children, while at the same time adults are capable of playing
a guiding role in their own learning. Malcolm Knowles's work helped lead
the way to North American discussions of the "facilitation" rather than the
"pedagogy" of adult learning and this formed a fundamental departure for
understandings of how adults learned. However, Knowles did not ad-

equately address collective learning, nor did he think critically about how social standpoints such as class, race and gender might affect learning practice. More importantly, Knowles took for granted that expert/novice relations still defined the learning process, as even adult learning required some more skilful other to play the role of facilitator.

In contrast, "self-directed learning" first conceptually developed by Allen Tough (1967; 1979; see Percy, Burton and Withnall, 1994) moved beyond the imputed need for a facilitator in adult learning. Adults regularly engaged in a wide array of learning projects, and many scholars have since examined the ways and means of these processes. However, Tough's original empirical studies, and even the vast majority that followed in its footsteps, have remained largely class-, gender- and race-blind. Power relationships, time, resource and energy constraints as mediated by major social divisions all play important roles in shaping adult learning and are not addressed in this conceptual tradition. While the few studies of working people's self-directed learning have found a similar incidence to the informal learning of middle-class adults, adult learners are generally treated exclusively as individual learners. Moreover, studies of self-directed learning miss the point that a great deal of learning cannot be discussed because it is tacit and/or must be performed, often with others, rather than reported individually.

Perspective transformation or transformative learning theory emerged in the late 1980s (Mezirow, 1991, 1994, 1996) and was more sensitive to the learning potential of critical events in the adult life course. Mezirow offered a critical view of communication and social difference. Building on the work of the Frankfurt School, and specifically the formulation of Jürgen Habermas, the theory of transformative learning centres on relations of communicative action which are said to be key to understanding learning that synthesizes the disparate and competing forms of rationality. Nevertheless, as Hart (1992) points out, specific social standpoints are not well recognized nor, as Newman (1994) observes, is the nitty-gritty world of material life, so that the possibility of the ideal communicative exchange is often severely limited.

In contrast to transformative learning, a more developed engagement with critical theory and Western Marxism more broadly can be found in the tradition of critical pedagogy (Corrigan, 1979; Giroux, 1983; Livingstone, 1987). Here we find a superior explication of the complexity of education and more general learning relations in advanced capitalist society. The writings of the Brazilian popular educator, Paulo Freire (1970), have inspired the development of this perspective; while writers such as Giroux and Corrigan, for example, have offered effective explorations of educational life emphasizing the reproduction of culture, subjectivity and human agency in capitalist society. The relevance of this particular set of approaches for our purposes lies in its commitment to a careful consideration of praxis from the standpoint of the oppressed themselves (Giroux, 1983; Livingstone, 1987). However, critical pedagogy's empirical focus has remained largely fixed on the domain of formalized education and rooted in relations of pedagogy. Though Freire himself was deeply committed to the notion of learning outside formal schooling, critical pedagogy in the North American context appears to have been only minimally concerned with actively researching and theorizing a broader conception of learning per se, a fact increasingly noticed by critical pedagogy theorists themselves.

Within the context of critical pedagogy, the work of Paulo Freire, and particularly the notion of "conscientization" deserves special attention. Developed in the early 1960s, Freire's concept of conscientization expressed an explicit commitment to understanding relations of learning as grounded in specific racialized, gendered, classed and regionalized standpoints of the third world. As Freire notes, the standpoint of the oppressed provides the basis (vis-à-vis "generative themes") upon which a critique of schooling, curriculum and social life ultimately rest. Learning is understood as rooted in social practice in the concrete world and what Freire and Macedo called the "*act* of knowledge" (1987: 52; our emphasis). Indeed, for us, this recognition of the importance of concrete, social practice for the learning process emphasizes that it is more than mere coincidence that Freire identifies Vygotsky, founder of cultural historical activity theory, as one of his early influences (Horton and Freire, 1990: 36). However, the concept of conscientization has clear limits for the type of analysis of working-class learning in this book. Although grounded by Freire in collective political action implicitly, the concept of conscientization itself does not offer a means of making visible the actual social nature of the learning processes of subordinate groups. This suggests a bias towards conscious goal-directed and cognitive dimensions of learning and social action. We are left with engaged description rather than a conceptual means to carry out systematic analysis. And, as with androgogy and critical pedagogy, conscientization seems to assume an "enlightened" other.

While these concepts and theoretical traditions each have something important to offer in terms of a grounded analysis of working-class learning, it is our claim that they represent perspectives with fundamental limitations as well. Part of the reason for these limitations, perhaps, has to do with the empirical work upon which each approach is based. In these terms, several recent attempts to understand learning in relation to work specifically have somewhat greater potential for our analysis. Focusing on employment relations, the "point of production," puts the class struggle front and centre, in effect demanding that the analyst make a choice to either recognize or dismiss the principle of competing social standpoints in the learning process. In short, in the paid workplace, conflict is "in one's face:" one must either turn away or actively engage with its contradictions. Authors such as Hart (1992, 1995) have been very insightful in this regard. In particular, the work of Foley (1999) provides an exemplary approach in terms of research sensitivity to workers' standpoints within the learning process. Foley's *Learning in Social Action* sets a standard that, in several ways, we seek to mirror in our application of the principles of a historical materialist approach to learning that moves beyond the core, interlocking themes of individualist/cognitivist-rationalist bias, universalist/ahistorical bias, and the bias of formalized learning.

Origins of Cultural Historical Activity Theory

The critique of adult learning theory above prefigures a call for an alternative. To best understand this alternative it is important to begin with first principles. It is therefore worthwhile considering the exercise that cultural historical activity theorist (CHAT) Yrjö Engeström (1987) undertook in looking back at some of the traditions most relevant for a full understanding of learning as historical, truly human, activity. At the heart of each of these traditions, according to Engeström, is their attempt to grapple with two key orienting principles: first, the notion of tool-mediation as a means of escaping the long-established Cartesian separation of mind and body; second, the need for an account of the dialectical nature of structure and human agency in practice. For Engeström, one relevant lineage that seeks to speak to these principles is to be found in the semiotics of C.S. Peirce (1931–35). He argued that, although Peirce produces a detailed treatment of the process of mediation between subject (or interpretant) and object vis-à-vis the sign, the mediational process is still restricted to its intentional, abstract-logical and linguistic dimensions. In turn, the lack of connection between symbolic mediation and the realm of concrete action inspires Engeström to conclude that Peirce's formulation loses much of its power to form a viable alternative to Cartesian approaches to learning and human activity. As such, the Peirce lineage displays many of the same biases of the adult learning theory we criticized in the previous section, specifically in terms of its possible individualist and cognitivist bias.

The second lineage Engeström outlines originates in the social psychology of George Herbert Mead (1934). In Mead's work, we see a type of tool-mediated process of meaning production, understood vis-à-vis the basic formulation of gesture/adjustive response/result. However, Engeström's critique of Mead suggests that, once again, we are faced with a limited conceptualization of human practice as a process of communication separated from the production of sensuous, material practice mediated by tools that may be material as well as symbolic. In Engeström's view, Mead's much narrower explorations of material dimensions of practice in *Mind, Self and Society* do not do enough to demonstrate the relationship between communication and the overall cultural and material structure of social life and social change.

For Engeström, as for us, the third and most promising lineage is to be found in the cultural-historical school of Soviet psychology and the work of the Russian psychologist, Lev Vygotsky (1978; 1994). This tradition emerged from an attempt to understand learning as a historical form of human activity with an explicit concern for the problematics of peasant and working-class groups. It was the work of Vygotsky and colleagues such as Luria and Leont'ev that gave rise to the development of CHAT. Based on key concepts of tool-mediation and what he called the "zone of proximal development" (ZPD), Vygotsky turned learning (as understood as explained above) on its head, in that the relations of thought and action, material practice, and hence social participation, were regarded as primary to cognition. An important feature of the development of this lineage, starting with Vygotsky, was the context of its birth. During the earliest phases of the attempted construction of the Soviet Union into a state socialist society there were a wide variety of open declarations and ideological supports for the social scientific valorization of the lives of previously oppressed classes. Having attended the famous Shaniavsky People's University (home of the leading anti-czarist scholars of pre-revolutionary Moscow), Vygotsky was heavily influenced by the works of Spinoza, Hegel and, especially, Engels and Marx (Blanck, 1993). Indeed Vygotsky's program was aimed at nothing less than a reorganization of psychology based on Marxist principles:

> It is my belief, based upon a dialectical materialist approach to the analysis of human history, that human behaviour differs qualitatively from animal behaviour to the extent that the adaptability had historical [elements] ... Naturalism in [bourgeois] historical analysis, according to Engels, manifests itself in the assumption that only nature affects human beings and only natural conditions determine historical development. The dialectical approach, while admitting the influence of nature on man [sic], asserts that man, in turn, affects nature and creates through his changes in nature new natural conditions for his social existence. This position is the keystone of our approach to the study and interpretation of man's higher psy-

chological functions and serves as the basis for the new methods of experi-
mentation and analysis that we advocate. (Vygotsky, 1978: 60–61)

In collaboration with Vygotsky, but under the guidance of Luria
(1976), related ethnographic and linguistic research conducted with workers also began to provide important insights into the interrelations between
modes of production and modes of thinking and learning. Following
Vygotsky's premature death in 1934 and an extended period of suppression, Vygotsky's work slowly re-emerged in the post–World War II period
and, in the last two decades, research in this tradition has seen a vigorous
growth in the West.

The core contribution of Vygotsky's work (1985: 46) lay in the argument that "learning" is a "socio-cultural" rather than simply a cognitive
phenomenon. It is mediated by symbolic as well as material tools and particular modes of social participation that allow first inter-subjective and
then higher order mental functions of cognition and self-directed practice.
That is, the concept of learning presupposes both a social and a historical
nature (Vygotsky, 1978: 88). In these terms, Vygotsky's notion of "turning" is vital. Turning or interiorization defines the process in which external social relations and socio-historical systems are transformed into the
internal mental actions, outcomes and embodied states we associate with
notions of knowledge and skill. Both in his own time and ours, researchers beginning from these basic, first principles have attempted to treat the
foundation of adult learning as cultural, historical and participatory rather
than individualized, internalized or universalized as dominant theories
would have it. But these are only first principles.

Activity Theory From Vygotsky to the Present

Building on the paradigmatic break in learning theory that Vygotsky made
possible, the development of a recognizable CHAT approach is generally
ascribed to his colleague A.N. Leont'ev. Leont'ev's work is typically considered the bridge between Vygotsky's formulation of first principles and
the development of modern theories of CHAT. In his texts, *Activity, Consciousness, and Personality* (1978) and, in particular, *Problems of the
Development of the Mind* (1981), Leont'ev outlines the central critiques of
psychology put forth by the cultural-historical school and pays special attention to the foundational role of Marx's critique of bourgeois philosophy. For Leont'ev, Vygotsky provided a truly revolutionary extension of
Marxist analysis (1978: 11) to encompass the means of understanding
learning as participation in social practice defined by dynamic transforma-
tions, change and interrelations within social systems. Leont'ev (1978:
10) defined activity as follows:

> The minimal meaningful context for understanding individual actions.... In all its varied forms, the activity of the human individual is a system set within a system of social relations ... The activity of individual people thus depends on their social position, the conditions that fall to their lot, and an accumulation of idiosyncratic, individual factors. Human activity is not a relation between a person and a society that confronts him.... [I]n a society a person does not simply find external conditions to which he must adapt his activity, but, rather, these very social conditions bear within themselves the motives and goals of his activity, its means and modes.

It is within this early framework that we see a perspective on adult learning that necessarily required that practice be historicized and contextualized broadly. Expanding on the basic concept, activity is associated with a series of more specific concepts that provide the analysis of activity with multiple layers. Leont'ev (1978: 173–4) offered an explanation of the need for these different levels of activity:

> An analysis leading to an actual disclosure of sense cannot be limited to superficial observation ... After all, from the process itself it is not evident what kind of process it is – action or activity. Often in order to explain this, active investigation is required: substantiating observation, hypothesis, effective verification. That to which the given process is directed may seem to be inducing it, embodying its motives; if this is so, then it is activity. But this same process may be induced by a completely different motive not at all coinciding with that to which it is directed as its results; then it is an action.... In spite of what it seems to be from the superficial point of view, this is a way that confirms the objectivity of its bases to a high degree inasmuch as this way leads to an understanding of the consciousness of man derived from life, from concrete beginnings, and not from the laws of consciousness of surrounding people, not from knowledge.

Leont'ev's conceptualization of activity moves Vygotsky's observations into a workable model of adult learning. It provides the means to begin to systematically analyze everyday learning, beyond references to conscious reflection, as something more than simply a shapeless flow of experience. Through it, learning practices come to be seen as specific forms of political and economic organization, understandable in terms of different (and potentially conflictual) social standpoints.

To summarize the relationship between Vygotsky and later developments in the tradition we can refer to what are often identified in the CHAT literature as "generational" phases of development. The first generation, characterized in the discussion of "first principles" in the preceding section, centres around the notion of human action and human learning as mediated by cultural tools understood in the context of the "zone of proximal development" (ZPD). ZPD is "the distance between the actual developmental level as determined by independent problem solving and the level of potential development as determined through problem solving under adult guidance or in collaboration with more capable peers" (Vygotsky, 1978: 86). Leont'ev's development of the concept of "activity", implicating in many ways the idea of the ZPD, is associated with the second generation of theory. Here the realization of the importance of the role of mediating cultural tools is elaborated to include broader social dimensions of practice.

The most recent generation of CHAT revolves around six key analytic principles. First, it maintains a central role for seminal concepts such as *tool-mediation*, which is inclusive of semiotic as well as material artefacts. The latter has been investigated by writers such as Billett (e.g., 2001) looking closely at work activity investigated in terms of "affordances." Second, the principle of analysis *vis-à-vis* the notion of the ZPD is necessary. The ZPD allows us to understand the bases for Leont'ev's later claim about the holistic nature of learning practice. The pertinence of ZPD to processes of working-class adult learning from mentors and with peers must be recognized. Third, contemporary CHAT stresses understanding *articulating activity systems*. This principle focuses on how activity systems are enmeshed within a "multi-dimensional network of [other] activity systems" that interact, support, destabilize and interpenetrate (Engeström, 1992: 13; see also Engeström, Engeström and Kärkkäinen, 1995). Our notion of three spheres of work and four spheres of learning is compatible with this principle. However, while these three principles appear in the literature as unproblematically accepted, there are others that are more unevenly understood, unevenly utilized, and disputed.

It is more or less accepted that the fourth principle, *historicity*, must be included in viable, contemporary CHAT analyses. In conventional terms, this principle says that activity systems are always seen to be part of a development trajectory. In less conventional terms however, this principle, rooted in the original work of Vygotsky, must also be seen as part of the Marxist tradition of dialectic analysis of a system "in motion" (Ollman, 1993). In Engeström's core theoretical contribution to date, *Learning by Expanding* (1987), he forefronts the role of Marxist dialectics, including the notion of "ascending from the abstract to the concrete," drawing on the work of Marxist philosopher Evald Il'yenkov (1982). Historicity, then, intimately relates to a fifth principle: that is, the role of systemic *contradictions* in the developmental trajectories of activity systems. Following Il'enkov's elaboration of the concept, Engeström (1987) goes on to outline how practice undergoes the type of change that defines the learning process. Thus, rooted in the internal contradictions of the activity system, an "expansive cycle," involving individuals questioning and eventually moving beyond legitimated forms of practice, is produced. New patterns of individual participation, and even new patterns of activity as a whole, ensue. The sixth and final principle that orients our analysis in this book is the concept of *alternative social standpoints*, which writers such as Engeström (1992) and Wertsch (1991) have discussed in terms of the "heterogeneity of expertise" or "multi-voiced-ness."

However, systemic contradiction and alternative standpoints are increasingly points of contention among CHAT theorists. For us, the most important point of contention revolves around the degree to which the historical materialist perspectives upon which Vygotsky began, are recognized and applied. Vygotsky and CHAT perspectives have entered mainstream intellectual traditions, albeit as some authors contend (e.g., Elhammoumi, 2000; Newman and Holzman, 1993; Rowlands, 2000; Sawchuk, 2003a), with the "domestication" of core historical materialist elements. Through our presentation in this book, we argue that central among these domesticated elements in need of reassertion, are critical notions of *working-class standpoints* on the one hand (including patriarchal, racist, ageist and ablist forms of oppression) and the historical/political economic dimensions of the concept of *contradiction*, as originally developed by Marx, on the other. In terms of the former, we emphasize that the recognition of alternative social standpoints, far from being a "relativistic" enterprise (cf. Rowlands, 2000), is at the heart of what Lukács (1971) characterized as objective, historical materialist analysis. Marx himself was clear on this matter. The analysis of social processes from the standpoint of subordinate, working-class groups was an ontological and epistemological point of departure, not simply a political commitment. Moreover, the notion of *contradictions* (which comes clearly into view when we recognize conflicting social standpoints) is both the motor of historical change (Marx) and definitive of the learning process (Vygotsky). Thus, the concrete realizations of racism, patriarchy, unemployment, poverty, and class struggle generally, are neither periodic nor exogenous to activity (cf. Engeström, 1987; Elhammoumi, 2000), but rather forms of contradiction fundamental to it.

CHAT From a Working-Class Standpoint

Our use of an activity theory perspective in this project has been developmental. We began with a very general appreciation of Vygotsky's writings and central concepts, as well as the intention to explore their relevance to more specific and largely ignored forms of adult working-class learning practices. In our overall analysis of working-class learning in the paid workplace as well as in the home and community, each of the above six CHAT principles of learning is pertinent. We refer throughout to mediation of practice vis-à-vis such "tools" as the formalized conventions of the collective agreement (e.g., auto workers), new forms of technology in the workplace (e.g., garment workers), and even the unique narratives of industrial conflict that workers develop themselves (e.g., chemical workers). Our understanding of the learning process implicitly assumes that learning emerges from patterns of social participation rather than strictly internalized, cognitive processes and in these terms we invoke the concept of activity and the principle of the ZPD: co-workers, predominantly through various informal networks, collaborate and construct a skills and knowledge "scaffold" for greater individual and collective knowledgeability. This often involves more senior workers informally guiding less experienced workers and sometimes collaborative experimenting. These tools, in themselves, are created out of specific divisions of labour and thus represent a concretized historicity; however, historicity also plays a key role in our analysis in the form of our explorations of firm and sectoral trajectories in each site chapter, as well as the more specific histories of labour process change that workers in different firms and departments face over time. Interlocking sets of subordinated social standpoints, including gender, race and age-based relations as well as class, orient our entire framework. When we underscore the forms of race and gender-based occupational segregation, the complexity of these interlocking standpoints becomes particularly clear. Likewise, we understand particular systems of participation, or activity, in terms of their articulation with others throughout the firm or across the sector, or even on the scale of a globalized economy (as in our discussion of the role of trade agreements in auto and garment sectors). Moreover, articulating systems of activity are in many ways the topic of inquiry in our chapter on family and community learning. Here we see how the three main spheres of work (paid, voluntary community and housework) are intertwined. This interaction has particular importance for our knowledge/power thesis, as learning in one sphere (e.g., volunteer work) may be a site of greater discretionary control or power, and at the same time may influence learning in other spheres, such as the paid workplace, where workers experience less control. Finally, the central contradictions of capitalism rooted in the commodity form (cf. Engeström, 1987), but more specifically in the forms of everyday class struggle, emerge in clear view in all site analyses. Forms of systemic contradiction, mediated by the differential powers of employers and organized workers, drive the emergence of particular patterns of exclusion in training; they also drive the patterns of informal networking that workers develop in response to regressive labour processes that ignore, deny, denigrate and de-legitimate worker skill and knowledge.

So, we therefore understand both the labour processes and learning processes that occur in and around the paid workplaces of advanced capitalist societies to be either driven primarily or strongly influenced by the dynamics of inter-firm competition, class struggles and continual modification of the forces of production. There has long been a substantial structural correspondence between the hierarchical and competitive relations of production and the organization of formal school systems to prepare and select students for entry into the production system. These production relations in their recently accelerated form compel corporate enterprise owners/top managers to increasingly recognize and attempt to use the knowledge and learning capacities of the workers they retain, while working people themselves continue to acquire further knowledge and certifications in order to live and work effectively in this market-driven, advanced technological society (see Livingstone, 1999a). The prevalent tendency in capitalist economic and social thought has been to reduce production relations to monetary exchanges between employers and individual wage earners, and to regard social relations generally in terms of self-interested consumption, or "possessive individualism." Marx amply demonstrated this optic to be an illusory version of social reality well over a century ago. Since then the growing division of labour within the capitalist production process has required increasing interdependency or cooperation among hired workers, increasing *collective labour*. As the production process has become more information-based than materials based, this has also required more *collective learning* by the engaged workforce. Nevertheless, as noted above, dominant educational thought remains preoccupied with individual learning motives and formal educational credentials while ignoring the informal and collective learning processes that working-class people perhaps rely upon most.

Simple reproduction theories of education that see learning practices in terms of a straightforward structural correspondence between the social relations of production and the relations of schooling miss the multidimensionality of both learning and work as well as the contradictory class relations that animate actual working-class practices in both areas (see Livingstone, 1999a). But the dominant influence of capitalist schooling on working people's learning dispositions must be acknowledged. In spite of systemic class biases against them, children from working-class origins with greater inherent intelligence sometimes do perform well individually in upper-middle-class–dominated classrooms. As more and more younger working-class people have stayed in school longer and credential upgrading has been increasingly required by prospective employers, advanced educational institutions have become somewhat more responsive to working-class experience. The more positively working-class youths have regarded this schooling, the more likely they have been to continue to participate in adult education courses within the same institutions, and the class gap in adult education has narrowed as Table 0.1 suggests. But the university remains dominated by those from more affluent class origins and the majority of people of working-class origins who manage to obtain a university degree now end up leaving the working-class. This is regarded as a matter of "natural" individual mobility in an inevitably hierarchical occupational order by people who blithely accept the absurdity that corporate executives could inherently be more than ten times as likely to benefit from a university education as those adults who remain in the working-class.

But capitalist dominated educational institutions have never managed to become dominant in relation to working-class informal learning practices which the CHAT approach and prior survey and anecdotal evidence (see Livingstone, 1999a, 1999b; Sawchuk, 2003a) suggest continue may prevail in the spaces and sites they still largely control – workplace peer relations, union membership relations and their own households. Informal working-class learning, conceived of and estimated in these terms, appears to display virtually the same frequency and strength of interest as that of corporate executives. While most working-class learning has historically remained hidden or ignored, both the previous empirical findings and the potential for more democratic information sharing by workers through widening technological communications networks (see Dyer-Witherford, 1999) suggest that this collective informal learning capacity could become a social force to be reckoned with. The fuller recognition, which studies like ours aim to provide, of the extent and richness of working-class informal learning – often under extraordinarily difficult material conditions – could enable more equitable recognition of working-class capacities generally. However, further development of the CHAT approach clearly requires integrative consideration of the strong influences of other mediating social relations besides simply class-based forms.

Patriarchial, Racist and Ageist Effects on Working-Class Learning

Capitalist societies are grounded in multiple oppressions, not just the general exploitation of workers. The most relevant for our purposes of comprehending working-class learning is patriarchal oppression which men have asserted through economic institutions that established their dominance in property relations and male breadwinner power as the main income earner; through political institutions in which men have legitimated these powers in legal and policy forms, and through patriarchal cultural forms that have attempted to naturalize male property rights, household head status and general superiority over women. Working-class women have remained relegated to primary household roles while increasingly taking up secondary wage earner statuses in poorer paid, less secure job ghettos. This has meant a double day of domestic and wage labour, which leaves much less free time for women to engage in discretionary activities such as learning projects. This double day is true for the vast majority of working-class women in our study. The patriarchal structure may also mean that women tend to give higher priority than men do to household learning related to these primary task responsibilities and have even less time to focus on or devote to paid work-related learning. While our main focus is on documenting work-related learning for both sexes in relation to paid wage work and secondly to household and community work, our case studies also trace the influences of patriarchal structures on these working-class learning processes.

Production/skilled trades differences among workers are likely to be closely related to both gender differences and opportunities for job-related learning. Tradesmen generally have more organized training and apprenticeship opportunities related to their jobs as well as more continuing discretionary control over the pace of their work and job-centred learning than production workers. Nearly all trades workers are still men. Production workers may be of either sex but women are more likely to have jobs generally regarded as less skilled, less well paid and temporary or part-time status with less general commitment from their employers, including less support for training programs for women. Our case studies will document clearly the divergent learning contexts of trades and production workers, and we note the gendered dimensions of these differences specifically.

Capitalist production relations are also grounded in racial and ethnic oppression. Capitalist employers may either foment or accept racial and ethnic discrimination within their work forces because it pays to keep workers divided against each other rather than united against their bosses. White workers may be receptive to racial stereotyping and scapegoating of visible minorities when greatly concerned about their own job insecurities or the future job prospects of their own children. However, progressive unions (beginning with those related to the Congress of Industrial Organizations [CIO] in the 1930s to 1940s and more recently in many unions espousing broad social movement perspectives), including union leaders who took part in this study, have played significant anti-racist educational roles to overcome such scapegoating and build solidarity among their ethnically diverse memberships. Visible minority workers, particularly recent immigrants, have often found organized education programs the most obvious route to try to overcome such racial barriers. But racial divisions among local workforces likely continue to represent a serious barrier to solidarity and action to build more equitable workplaces. Visible minority women workers may face the most difficult barriers of all, and our case studies show in detail how these divisions operate in some sectors.

Capitalism is an ageist system as well as patriarchal and racist. In addition to being disproportionately white and male, the large and small employers, managers and professionals who dominate the earnings and control structures of capitalist enterprises are also predominantly middle-aged. The historical tendencies of capitalist dynamics towards capital intensification and automation have led, with generalized commodification of work in advanced industrial market economies, to the decline of decent paid jobs, usually accompanied by artificial prolongation of schooling and increasing chronic unemployment/underemployment for youths as well as earlier retirement and employment discouragements for older workers. Thus, younger people may be generally the most motivated to engage in any forms of organized learning or certification that might enhance their job chances in an increasingly credential-laden labour market, as well as to engage in more informal learning to cope with the general transition to adult life. Older workers may respond to alienating job conditions by relying heavily on their cumulative experience and reduce their active and creative engagement in learning related to the production process. In general, if you are middle-aged, and especially if you are also white, male and upper-middle class, your job-related learning activities are more likely to be voluntary and of middling frequency between naive and desperate youths and resigned elders. Similar generational patterns of learning practices are likely found within the working-class as well. Ageist dimensions of learning, in the form of differential credentialization and attitudes towards work and learning, are operative in most case studies.

Our application of a CHAT approach to adult learning based on these considerations remains very preliminary in the following analyses. Our basic theoretical objective is to begin to demonstrate the fruitfulness of CHAT-oriented explanations of actual learning practices amongst organized working-class adults in advanced capitalist settings.[8]

Concluding Remarks

Adult learning is an embodied dimension of ongoing cultural material life. The real richness and complexity of actual working-class learning practices in contemporary everyday settings inherently involves often oppositional gender, race and generational relations in interaction with class-based practices. Working-class learning is inseparable from the diverse historical conditions of its production in local settings. It includes practices and experiences that are not typically framed as learning. This expanded conception of learning rejects the pervasive notion of learning as a moment of internalization, of cognitive processing by the universalized free-floating individual and of transference of "knowledge/skill" from some type of expert/pedagogue. Instead, we locate "learning" in the process of participation in the creation and reproduction of systems of activity in the sense developed in the cultural historical activity theory of learning. These activity systems include not only capitalist-dominated schools and labour markets, but some worker-controlled social institutions such as trade unions, the household and the neighbourhood. These processes of participation are subject to limits and pressures that shape but do not determine actual learning and its outcomes. As broad as this approach is, it permits an empirical research program that can provide meaningful and relevant accounts of working-class learning which go beyond the narrow hegemonic scope of cultural capital theory and modernist adult-learning theories and which, rather than reasserting dominant institutional forms of schooling, can contribute to the challenging of these forms by subordinated groups themselves.

The evident historical instances of creative cultural practices and preliminary evidence of extensive informal learning presented here from within worker-organized settings should serve as an antidote to the cultural capital bias for any researcher or teacher who cares to look. The working-class culture expressed in these settings should not be romanticized; as we will see, like any culture, it remains full of inconsistencies, unevenness and reactionary aspects. But, in spite of much academic opinion to the contrary, it is from these most concentrated and independent sites of collective expression of this class culture that struggles against bourgeois cultural hegemony and for democracy and economic justice are likely to be sustained. This is the cultural process that Raymond Williams (1975: 241) aptly called the "long revolution":

> I believe in the necessary economic struggle of the organized working-class. I believe that this is still the most creative activity in our society. But I know that there is a profoundly necessary job to do in relation to

the processes of cultural hegemony itself. I believe that the system of meanings and values which a capitalist society has generated has to be defeated in general and in detail by the most sustained kinds of intellectual and educational work.

As Williams and other working-class intellectuals know from their own experience, a definitive feature of working-class learning is the emergence from collective social practices of critical insights into the limits of competitive individualism and a sense of alternatives. These types of critical insights continually emerge from capitalist employment settings whenever workers are able to share their experiences. During one of our group discussions, Tom, a chemical worker, described the role of participation in the labour union for learning about the political economic context:

> The union's role as I see it is to highlight what shared knowledge we actually do have and how we attain it, how we actually do learn things. I'll give you a quick example. Four people in this room learned about health and safety the hard way. They learned about workers' compensation the hard way. Only through their experience. They never went to any union course – they learnt it when the employer fucked them and then they had the time to sit down and say, "Why'd they do that to me – after all I've given them." And that's the best, unfortunately it's the hardest as well, the best experience a worker can get because it cuts through all the nonsense because it hits you directly, it gives you time to think and to read and ask questions and start understanding what it's all about.

Part II of this book reports the findings of some of the detailed work necessary to appreciate the often hidden critical and creative capacities of working-class learners.

Notes

1. See, for example, Perry Anderson's (1976) account of the twentieth century development of Western Marxism.

2. For a recent study which reviews various academic Marxist and Weberian theories of class structure and class consciousness (including Bourdieu's), identifies mediated forms of class, gender and race consciousness, and examines their current expressions in a local attitude survey study, see Livingstone and Mangan (1996).

3. For a fuller critique of consequent idealist tendencies and a revised materialist theory of group consciousness grounded in case studies with industrial workers and their partners, see Seccombe and Livingstone (1999).

4. From Bourdieu's wide array of writings, the most pertinent mature works are Bourdieu (1984, 1991). In Bernstein's narrower corpus, see Bernstein (1990, 1996). For some of the most extensive and insightful critiques to date, see Fowler (1997) and Swartz (1997) on Bourdieu, and Sadovnik (1995) on Bernstein.

5. Bohman (1999: 135) makes a similar critique of Bourdieu's general conception of human agency:

   Actors are, in effect, "cultural dupes" to their habitus as they were judgmental dupes to Parsonian norms.... Bourdieu needs to be clearer that even shared means are subject to constant interpretation and reinterpretations, often in ways that contest current identities and practices.

6. For an insightful account of the development and demise of the proletkult movement and of the struggles between the popular and vanguard forces associated with Bogdanov and Lenin, see Sochor (1988).

7. See especially their own accounts of their adult education experiences in McIlroy and Westwood (1993) and Thompson (1968).

8. For another recent example of a research project applying an activity theory research perspective to worker training programs from organized workers' standpoints, see Worthen (2001).

PART II

CASE STUDIES

CHAPTER 3

Auto Workers: Lean Manufacturing and Rich Learning

co-authored with Reuben Roth

Pete Jones is an assembly-line worker in the auto plant. He is in his late 20s and has worked in the auto industry most of his adult life. Pete is very active in his union local and his parents were both union members in other industries. His family is of European ancestry. He was also an active and creative youth, but some of his teachers saw him as just another working-class child with limited prospects.

> *I was really hyper when I was younger, but just very curious. In school, I was a bad student, a horrible student.... It was always a struggle.... In grade eight, the guidance counsellor told me that I should take basic English because I was a failure at academic English, mainly because my spelling was bad.... When I got to high school, all of a sudden they assume you could read and once I had to read books, like Shakespeare or whatever, guess what? They're interesting. I could understand the story and it was all very logical ... it was easy.... Then I had a math teacher who said why don't you just skip if you have no interest in participating.... As soon as I could get out of high school, I did.... I took night classes for a while but pretty soon everything went under.*

In contrast, Pete's informal adult learning activities are much more coherent and sustained, and clearly linked to his own social and political interests:

> *I'll spend time learning anything to be better; music, computers, but the union is the big passion for me, like I feel like that's the thing where I can have the most opportunity to do things. For me, it's like that's what I want to do, you know.*

Pete has taken numerous union-sponsored courses related to employment benefits and job assessment as well as social issues. But most of his learning, and especially his union-related learning, is done in informal collective settings by drawing on the knowledge of more experienced colleagues:

> *Most of my learning is done with other workers. Whether it's fighting [against an oppressive] production standard or whether it's a problem in a certain area, if it's an [employment] insurance problem or whatever, you know. You seek out the advice of somebody else and there's always something to learn about a different part of the collective agreement or a different approach to how to handle a certain job or, you know, a specific problem that somebody's come to you with. So talking to other workers is a good way to do it.*

Pete composes and plays music to express many of his thoughts on a variety of political and social issues, drawing inspiration both from his own experiences and other cultural sources that resonate with these experiences.

Pete's creative impulses in his music and his daily activities with his workmates are frequently driven by his sense of the social injustices in capitalist society. He has developed a very concrete understanding of the central exploitative relations of capitalist production systems:

> *Management is screwing my people day after day ... ruining our lives, screwing up our jobs, eliminating work, adding the remaining work onto everybody else. We're working harder than we ever have before and these [management] guys can sit and laze around. ... They're overpaid, we actually earn our money and more profits for the company. We're assets, they're expenses. ...We had a guy fired recently for breaking a piece of equipment. A door got jammed and he got fired for restricting throughput because he couldn't work it. If the [company's] recourse is to fire somebody like that for having a broken piece of equipment, well, what happens if the company breaks my "equipment," my hands, say with carpal tunnel syndrome which is no longer listed as a workplace accident [under Ontario's compensation regulations]? You're going to be thrown on the scrap heap.*

Introduction

Automobile production is the prototypical twentieth-century industry. Production has been carried out through mass assembly processes in very large factories by large concentrations of unionized male workers, whose collective strength has allowed them to negotiate a relatively high "family wage" and various benefits for enduring routinized and tedious assembly labours. This "Fordist" production regime became widespread as the mass production techniques that originated in the industry were copied in other sectors from manufacturing to fast food outlets and government offices (e.g., Rinehart, 1997). But the old Fordist wage bargain has come under increasing pressure with heightened competition and further automation in the global auto industry. Since the 1980s, the auto industry has shifted towards more "lean production" techniques. In 1995 General Motors Canada's (GMC) past President Maureen Kempston Darkes proclaimed: "In response to competitive pressures and global events General Motors has embarked on a restructuring of its operations to reduce cost, improve quality and bring products to market that reflect state-of-the-art design and technology. In this regard, our Canadian car assembly plants have been leaders in the implementation of lean manufacturing processes, to eliminate waste and increase productivity in our operations." As described by Steve, a bodyshop worker at General Motors' Oshawa plant:

> There's been a continual move toward this "lean and mean" Japanese style of management ...They'll take a group of twenty people and they'll arbitrarily take one out and shift work – a little bit here, a little bit there. They'll look for the trouble spots that can't keep up and address them and address them until they've fixed it. Now they have nineteen workers working. Then down the road, they'll say, "Okay. Everything's running fine. Hasn't been a problem now for a month. Yank another one out. Take that work and shift it around." They do. Suddenly there's problems in the system, "have to catch up here, certain guys aren't keeping up." They'll rearrange it. Eventually, eighteen will be doing the work. And this is how they've been producing [reducing] the amount of workforce in the plant.

GMC's version of lean manufacturing is officially known as "syncronous manufacturing." This system has been persistently attributed in company training sessions to Japanese auto manufacturers, always depicted by upper management as "ahead of the Big Three" North American auto companies in terms of production efficiency and quality. Lean manufacturing is designed as a delicate balance between the stresses of production placed upon workers and the constant threat of parts shortages. This deliberate "fragility" is a key built-in feature of this form of "remaking" the workplace; it acts as an oversensitive trigger that alerts manufacturers to weak links in the production chain. Womack, Jones and Roos (1990: 102–103) have written that:

> ... lean production is fragile. Mass production is designed with buffers everywhere – extra inventory, extra space, extra workers – in order to make it function. Even when parts don't arrive on time or many workers call in sick or other workers fail to detect a problem before the product is

mass produced, the system still runs. However, to make lean production with no slack – no safety net – work at all, it is essential that every worker try *very* hard. Simply going through the motions of mass production with one's head down and mind elsewhere quickly leads to disaster with lean production.

The fragile character of lean production, intended to reveal weaknesses in the production chain, places pressure directly on the shoulders of assembly workers. The "buffers" referred to by Womack, Jones and Roos are both the manufacturer's and assembly worker's built-up stock of parts – a longstanding practice which allows for brief respites from the rigours of production. These buffers are built up at great sacrifice by workers who forego a steady work pace in order to speed up without pause. The reward of a few minutes' respite is all they have to look forward to. Thus, lean production relies on an industrial workforce that does not fall sick, does not tire, suffers no industrial accidents or injury and has an unremitting capacity for working at full "intensity" well over forty hours per week (overtime is a key feature – and failing – of lean production plants). In Lazonick's (1991) terms, this is predominantly an adaptive management regime in its reliance on intensified routines and tighter regulation of labour. As Rinehart, Huxley and Robertson (1997: 26) observe: "Lean production's proponents define it as a flexible system. It is flexible both in terms of product – the ability to change quantity easily and to produce different versions of the product – and in terms of the 'adjustment and rescheduling of human resources' to match fluctuation in production quotas."

The most advanced employer of lean manufacturing among Canadian auto manufacturers is GM's management. As Lewchuk, Roberts and McDonald (1996: 5) note:

> GM management aggressively moved to reduce staffing levels and maximize the use of available time. GM documents openly referred to their drive to implement Lean Manufacturing. They promoted the use of Andon cords [pull cords located at every operator's station which are used to stop the line when defects are found] for process control, pull systems to discipline the delivery of parts, material supermarkets, material card pull systems, visual line balancing process, ... Just-In-Time Production, flexible work cells, and best people practices. The latter are supposed to involve employees in establishing the best practices for an operation. What does this mean? To GM it means ensuring that actual operating time divided by the number of job cycles completed should equal planned cycle time. It means ... less time for moving to get parts, minimizing operator movements, and less time to rest and talk to your fellow worker. It means stand-

ardizing jobs, i.e. beginning and completing work within the physical work envelope, and performing work elements in the same sequence every time.

Lewchuk *et al.* found that GM workers directly involved in the assembly process were 30 percent to 40 percent more likely than workers at other Canadian auto companies to report that their workload was excessive. The increasing interdependence and reduction of buffers within the production process do permit some innovative worker responses. Through the use of "just-in-time" production methods (see Wells, 1987) the operations of GM and its auto parts production facilities are so closely interrelated that the three-week strike at General Motors' Canadian assembly plants in the fall of 1996 shut down dozens of GM's operations across North America. Conversely, a March 1996 U.S. brake plant strike in Michigan idled 175,000 GM workers across North America, including the entire Canadian GM workforce at four production and four parts facilities in four Canadian cities (van Alphen, 1996: C3). This underlines the central fact that the auto industry has, since the mid-1980s, used a tightly integrated system of production with a hair-trigger sensitivity to delays in production. Some might add that, given this system of production, auto assembly workers potentially wield more power over the production process than ever before. However, this relative economic might is not reflected in the current general atmosphere of employment instability (and the reality of downsizing and shifting automotive jobs to low-wage countries within the manufacturing sector). Thus autoworkers, despite their *relative* productive power, are understandably apprehensive about their future employment prospects. Lewchuk *et al.* (1996: 11) found that 72% of auto workers surveyed in nine Canadian auto assembly plants "reported concern about losing their jobs within the next three years."

According to Lewchuk *et al. 's* (1996) extensive survey of working conditions and attitudes at Canadian auto plants, the majority of workers reported that they could not alter the pace of work, nor could they talk to fellow workers outside of scheduled breaks. Additionally, workers even had a difficult time finding a temporary replacement worker (a "relief man") to allow a trip to the washroom. The study found that auto workers had to work at top speed to keep pace with the assembly line – building up a surplus supply of parts was no longer the practice. GM workers at all Canadian sites reported particularly low levels of autonomy and control, levels 30 percent lower than those found at other Canadian auto assembly plants (1996: 11).

Employment in the Canadian auto industry has remained relatively constant at around 55,000 since 1985 while output has grown significantly as a result of productivity increases. There has been an ample labour supply, with the exception of some skilled trades shortages with imminent retirements. New technologies diffuse rapidly from foreign companies and Canadian auto plants have continued to enjoy competitive advantage over U.S. plants through lower labour costs and socialized benefits like Medicare (Human Resources Development Canada, 2002). While GM, as the world's largest auto maker celebrates its emergence as "a far leaner, more flexible company" and "winners in the marketplace" (Weber, 2003), there is now significant global overcapacity that threatens additional plant closures in Canada[1] as elsewhere.

The General Motors Site in Oshawa

While the modern Canadian economy was founded on the extraction and export of staple commodities (furs, fish, wheat, metals), the auto industry has probably been the most strategically important single industry in the post-World War II era. In the mid-1990s, the auto industry provided almost 500,000 direct and indirect jobs (about 4 percent of total Canadian employment) and accounted for approximately 7 percent of Canada's GDP (Kempston Darkes, 1995). While free trade agreements with the United States and Mexico and the demise of the Autopact, which guaranteed privileged access to U.S. markets, now threaten the future sustainability of this wholly foreign-owned industry, auto production remains vital to the economic health of the country.

The GM site in Oshawa is the largest automotive assembly complex in North America and includes two car plants, a truck plant, a fabrication plant and a battery plant. Management calls it the "Autoplex", a term that is generally disliked by workers. GM is the largest employer in the city of Oshawa, located just 50 kilometers east of Toronto, and with a population of 146,000. Oshawa was the home of the McLaughlin Carriage Company which was sold to U.S. General Motors in 1919, signalling to some the death-knell of the Canadian-designed, Canadian-made automobiles (Robertson, 1995). At the beginning of the twenty-first century, Oshawa continues to be dominated by the auto industry. This city has been an area of *relatively* stable job security. The existence of a strong trade union is at the core of Oshawa's historic sense of security and has wide community effects. Oshawa and Windsor, another major auto production city in Ontario, have the highest median family incomes in Canada (Statistics Canada, 10 August, 2001).

The Canadian Auto Workers Union (CAW) is the largest industrial union in the country with over 215,000 members in Canada. CAW Local 222 in Oshawa is comprised of workers at the GM car and truck assembly plants located in Oshawa and a dozen other smaller units, totalling over 20,000 members in 1996 and making it one of the largest union locals in the country. The largest of these is the GM auto complex, with around 15,000 members in 1996 and 11,500 in 2002. GM Oshawa experienced a drop of almost 4,000 employees between 1996 and 2002, despite the addition of over 1,000 workers in August 2002 to staff the only third shift in a North American car plant.

Our study was conducted with GM members of Local 222. The GM unionized workforce has historically been almost exclusively male and white. After the 1993 "absorption" of laid-off workforce from GM's nearby Scarborough, Ontario van plant, many women entered Oshawa's assembly plants for the first time since World War II. They were joined by a small influx of visible minorities for the first time in Oshawa's history. But according to Lewchuk et al. (1996), the Oshawa car plant workforce remained 88 percent male when our study was conducted. This has been an ageing workforce with an average age of about forty-two and an average seniority of around fourteen years in 1996, increasing to forty-eight years average age and twenty years seniority by 2002. These figures reflect the fact that there was virtually no hiring during this period. There are no readily available estimates of the formal schooling levels of 222 workers, but the proportion of Oshawa residents without a high school certificate is above the national average (Statistics Canada, 1998) and at least until the mid-1970s GM workers had generally left school before completion to enter the plant. More recent hires are required to have at least a high school diploma. The vast majority of the 222 workforce are production workers while according to the local, skilled trades employees comprise approximately 17 percent of the plant population (Goggan, 2003).

In all, fourteen interviews were conducted with local 222 members. These included one woman worker, one worker of colour and two skilled tradesmen. But most of those interviewed reflected the dominant profile of 222. They were white, male production workers. Most had high school education or less, were middle aged and had average seniority of over fifteen years. It is not a representative sample but it is broadly reflective of the basic demographic profile of the plant.

The CAW and Local 222

Economic security for Oshawa auto workers was won through historic struggles, most notably the fight to join the United Auto Workers (UAW) union in 1937. The Oshawa strike of 1937 enjoyed the support of the entire city, including the mayor and local business owners, as well as two provincial cabinet ministers who resigned as a result of their vocal support

(Abella, 1975:1). According to Yates (1993), the 1985 separation between the UAW and the CAW can be attributed crucially to the democratic organizational structure of the then-Canadian Region of the UAW, as well as a more class-based, collective identity amongst its members and its anti-concessionary economic strategic direction. In contrast to the UAW, which provides few intermediary structures, a democratically elected Canadian Council is a fundamental internal mechanism that gives a voice and vote to elected rank-and-file delegates. This body both debates and decides the direction of the national leadership, and played a pivotal role during the UAW-CAW split. More recently, rank-and-file members responded to lean production measures in October 1996 with a GM strike and subsequent plant occupation. Local 222 still appears to be a fertile site of oppositional working-class struggle. As Greg, a plant operator, told us:

> I think people are getting tired of being threatened. I mean GM's mentality is, 'we keep threatening them enough, you take a little bit here and a little bit there, then they'll give a lot more here.' Well, people are fed up with it. I mean it's great to say "oh … I have a job." But … what cost are you going to give for your job?

Local 222 members are represented by numerous elected union representatives: full-time shop-floor representatives (committee persons, district committee persons and plant chairpersons) whose task it is to police the negotiated union-company contractual agreements. These shop-floor representatives are assisted by a team of service representatives who assist workers in navigating the various negotiated benefits (including sickness and accident insurance, life insurance, a dental plan, legal assistance benefits, an extensive drug plan, a short work week benefit that guarantees a full forty hours-worth of wages during work shortages, eyeglass, prosthetics and hearing aid benefit, Supplemental Unemployment Insurance Benefits [SUB], vacation pay, one Scheduled Paid Allowance [SPA] week per employee per year and one of the most envied private-sector pension plans in Canada). In a traditional democratic trade union practice, these groups and the local administrative body (the executive board) are elected from the shop floor by the entire membership body. In 1996 the typical GM assembler's wage was about $23 an hour – a relatively high rate for an "unskilled" blue-collar worker – plus an unusually generous benefits package the likes of which was almost unknown outside of the "Big Three" automakers.

Local 222's two longstanding caucuses, "The Autoworkers" and "The Democrats"[2] have existed in Oshawa in some form since the mid-1940s. In 1996 they were largely organizational groupings of like-minded and ambitious trade unionists, rather than formations based on an ideological core. Of course, this is possibly in large part due to the democratic nature of the CAW and the trade union movement. Caucus structures mimic those of the union local, with annual dues, an elected chairperson and executive, and internal caucus elections to decide which candidate will carry the caucus banner at local election time. In short, caucuses at Local 222 are highly structured groups. During elections the two major caucuses at Local 222 formed electoral "slates" of candidates who campaign under their respective caucus banner. Historically the Democrats, originally christened the "Democratic Right-Wing," were formed to combat the influence of the Communist Party of Canada (see Yates, 1993). More recently, during the 1985 split from the Detroit-based UAW, the Democrats and Autoworkers were divided over whether to forge a Canadian union or remain with the U.S. parent union (White, 1987: 325–328).

Local 222's caucuses are active mainly during in-plant union elections, but their influence has pervaded the union's day-to-day life. Their intense rivalries have often been acrimonious. However, all caucuses have had to respond with a united voice to the growing anxiety over industry restructuring and the threatened loss of jobs. Although there have been strong feelings about Local 222's two major caucuses, and groupings continue to shift, the existence of the caucuses is a testament to the democratic traditions within this union local. Caucuses fight for power because there is a measure of power worth fighting for. Moreover, they are an effective means of "spreading the word" efficiently and relatively inexpensively in a widespread community of workers. Finally, caucuses serve as a vehicle with which to train new cadre and move them smoothly up the ranks. In short, at Local 222 caucuses have become institutionalized democratic vehicles that serve as an effective training ground for union activists.

Tightening Labour Markets and Worker Rebellion

Our first interviews took place shortly after a massive January 1995 lineup of GM job applicants in Pickering, Ontario (twenty minutes west of Oshawa). Applicants responded to GM advertisements in what were previously unimaginable numbers. For two days news crews helicoptered above a sea of approximately 26,000 job hopefuls during two of the coldest January days in many years. This event was noted in some way by almost every respondent interviewed (indeed, it was noted by workers in our other research sites as well). Harry, an assembler, summarized his own in-plant experience and the almost daily use of the "Pickering army of the unemployed" threat by shop-floor management to intimidate workers who fall out of line: "I mean in our times right now … with General Motors, you say something [they] say 'Hey, you don't like it there's 30,000 people were lined up and freeze their butt to get in here.' *'You can't do it, somebody else can do it.'"*

In light of what they viewed as a deliberate, manipulative and inflammatory act, respondents keenly sensed their dispensability and feelings of job insecurity. This happened during a tense lead up to what was viewed as an inevitable strike against GM in the fall of 1996. The themes of anger, job insecurity and a belief that GM was not dealing fairly with workers, were perhaps the most persistent during these interviews. At the same time, workers also have contradictory feelings about their power, particularly during pivotal, collective events such as strikes and sitdowns. The atmosphere of unease, reinforced by the memory of thousands of eager applicants ready to assume GM jobs and displace seniority workers, often made for a resentful and angry group of respondents. Tim, a body shop operator, was agitated enough to make the claim that GM planned to eliminate the CAW from their Canadian plants. He cited a cluster of North American auto plant closings and the enormous two-day lineup as a signal of certain conspiracy:

> [T]here's so many things that have gone wrong lately that I have to question why, and I still ... haven't found a reason for it. Twenty-five thousand applications were taken in Pickering, General Motors ... around a time when plants were going to be idled or closed. It was obvious that there would be a massive pool to draw employment from, and yet they took 25,000 applications in Pickering. I have to question why.

In addition to the "Pickering lineup," GM unintentionally fed workers' anger when the corporation distributed black coffee mugs to all of their unionized, hourly assembly workers in celebration of GM's announced record annual profit of $6.9 billion (U.S.). The mugs, imprinted with the inscription "On Track ... In the Black," (a reference to GM's new found fiscal solvency) were in celebration of: "General Motors of Canada Ltd. [which] broke the record for annual profit by a Canadian corporation with a $1.391-billion tally that helped the parent company set a record high of its own" (Keenan, 1996: B1). GM Canada's contribution to its parent corporation's coffers was greatly out of proportion to the size of its market or workforce. The in-plant reaction was furious and the "gift" only added fuel to the fire. Unfortunately the intended effect was lost on Oshawa's workforce who calculated the "true" cost of the now-famous coffee mug. As Harry put it:

> Everybody say[s] 'Oh, that's a $25,000 cup!' If you figure it out, that's what it is. Well, you take the $6.2 billion [profit] and you split it into the

80

[number of] people who get the cup, that's how much it is.... They want [to] insult me? Go ahead. Insult me. But don't insult my intelligence.... The reaction was ... everybody was pissed.... Everybody ... smashed them [mugs] ... just outside of the parking lot.

While smashing a coffee mug may not be a revolutionary act, it is at least indicative of a general feeling of being "ripped off." The cup smashing affair, the three week fall 1996 strike and the ensuing GM North plant occupation are elements of a culture of resistance within this working-class community that are comparable to the cultural forms and sentiments that sensitive ethnographers have previously found expressed both within other working-class communities and among working-class children in school. These sentiments contributed to the extremely strong support from Local 222 members during the fall 1996 strike where issues of job outsourcing and mandatory overtime were the principal concerns. The feeling of being "ripped off" probably fostered the GM plant occupation, which was led primarily by skilled trades workers and took place in direct reaction to GM's threat to remove crucial parts dies from the plant.[3]

As soon as word of the plant occupation spread, Local 222 members from across Oshawa flocked to the North plant gates. During the course of the day, many striking workers jumped over the plant gate to join their friends inside. An act which may baffle outsiders, this action remains indicative of a unified common culture and shared economic understanding. In this case, workers won their demands and returned home that evening. The strike was considered a significant union victory and the North Plant takeover was the pivotal turning point. The North Plant occupation illustrates that there is a world of difference between the comfortable, established society outside and this working-class community with shared, deeply held values which require that, on occasion, its members step outside of society's comfort zone.

As our interviews generally suggest, Local 222 is one of the most concentrated and well-organized working-class communities in the country. At this particular union local, and more generally within the CAW, there is probably more extensive engagement of political education than in most other Canadian unions.

Working in the Auto Plant: "It doesn't take a genius"

The separation of intellectual and manual labour is viewed by 222's assembly workers as a chief distinguishing characteristic of their lives and occupations. For example, Harry describes his job thus:

My job! Well, I build doors. That's what it is. You just put parts in the machine and I crank them out and that's it.... Just put parts in the machine every day.... And what you do is that you put parts in the machine, you assemble them and you assemble the door. And the machine does the rest. You just pick up parts and you feed them through the machine you can build the ... door. That's it.... It doesn't take a genius to learn it. I mean, you know, they'll tell you how to put the part in and within an hour, you know how to do it.

Similar statements are often heard on the Oshawa assembly line; workers readily acknowledge that while they earn decent pay, their jobs involve little skill and satisfaction. Some learn their job within an hour while others do so within a day. Barney, another assembler, says:

[I] drill three holes – two small holes and a big hole ... I was trained ... by the guy that did it before me. Before he left, he gave me three days training.... Plus he got a guy on the other side on the line that does basically the same thing so you're working with another guy, so they sort of get a routine to work together.... it's pretty repetitive. You can learn it in about a half a day.... But to get your speed up, it takes a good three days.

Workers who learn to do their co-worker's job in addition to their own will be able to take an unauthorized break from their job while their linemate performs the functions of both jobs. This is referred to as "doubling up" and allows two workers a break from the routine of the line, as one worker "spells off" another. While one worker relaxes, reads or visits friends, his partner does the work of two, and vice versa. Supervisors often turn a blind eye to the practice of doubling up, realizing that workers regulate themselves and are unlikely to jeopardize their "free time" with substandard work. Lean production methods clearly threaten this discretionary time which is often needed for physical and mental recuperation.

Workers at 222, as in most other assembly line production systems are often unwilling to reveal any job-related shortcuts they have learned by doubling up, lest these are "stolen" by management and integrated into leaner production and more job loss. This is the principle behind the Japanese "Kaizen" model of organizing the workplace (see Parker and Slaughter, 1994). Bennie in the paint department speaks of the reluctance of lineworkers to do "their best" when their best will undoubtably be taken without any thanks:

I was shown how to do [my job] by a fellow ... I guess his concern was having the job done right, so he showed me easier ways to do things ... secrets to applying the tape correctly ... I'm lucky I pick up on things quickly, eh? Mechanically inclined, I've always been lucky that way, sometimes the corporation takes advantage of it, and they figure you can do everything ... they'll expect you to do everything. And if you say 'I can't do this' they get all cheesed off or in a huff because they expect you to be able to do everything ...

An example of stolen labour can be found in Harry's experience with the extensive computer training he received at community college. Although he attempted to leverage his knowledge of computers towards obtaining more fulfilling employment at GM, he found his skills used, but not acknowledged, by GM management. He describes an incident during the initial pilot stage of a new auto assembly process where his knowledge of computer spreadsheets was first recognized by a group leader as unique:

> They found out that I had computer knowledge. My group leader ... gave me a stack of papers, he said "you know, I only need these two pages, can you just print them for me?" So I just created a spreadsheet, printed them out [on] only one page ... he went to the meeting ... [where] everybody had this stack [of paper] ... and everybody said "where'd you get it?" "I have a friend who does it, he's on the floor, he's working with us." ... they said "Well, we need somebody who can do all this work." ... during the pilot I measured all these cars in a specific area.... gathered all that information, put it on a spreadsheet, charts, and by the end of the pilot had it ready for the engineer to hand it in for presentation. The engineer he took my name [off] and he put his name on it.

After this bitter experience – and several more like it – the respondent decided to no longer share his knowledge of computers with others at GM, although he does acknowledge that he gratefully receives several hours per week away from his assembly job in order to "help" his supervisor keep up-to-date records of overtime on his computer – a task his supervisor should actually perform. This respondent's thorough familiarity with computers has been exploited many times by his supervisors, with no (monetary) recognition, or promotion for his efforts. Since his is not a skill that is reimbursed by management, he has refocused his efforts outside GM and he now feels that his current attempts towards advanced training in the field of computers may eventually provide a "soft landing" in the event of a lay off. In other words, he has given up the possibility that GM might one day reward him for his knowledge and proffer him employment commensurate with this knowledge. After several attempts at advancement within GM, he acknowledges that his skills are "wasted" there.

Assembly line workers battle managers and supervisors in a daily attempt to rebalance an unequal power relationship. At times workers simply rebel out of a feeling of sheer boredom, a response to the repetition and lunacy of the assembly line. Life as an auto assembler verges on madness for many. Milkman (1997: 43) observes in her study of a GM factory in

California: "the metaphor of imprisonment was central to the self-conception of most GM workers ... because they hated their jobs so intensely." In this alienating atmosphere, an escape is needed but no other line options are more appealing for most assembly workers. As auto worker Greg says in candid response to whether there was another job at GM he might prefer: "Yeah, *retirement,* I got 'x' years, 'y' months and 'z' days remaining."

The contrast in the skilled trades workers' accounts of their jobs is quite stark. Gerhard is a fully certified millwright who works in maintenance through the plant. He describes his work with some relish: "Whenever something's going on [like a breakdown in the production line], Central Maintenance has to go. So it's quite interesting.... I love work, I never hardly make a break (laughs) I know guys probably think I ... don't play with a full deck, eh? ... I do enjoy it, right." His views on assembly work are even more harsh than those of assemblers themselves:

> On the line I would not feel the same. To be on the line day after day after day. I think this is punishment (laughs).... I think it must be awful. Be nobody, be like an ant who ... takes a leaf, cuts the leaf off and then carries it down in a hole. It's really ... not a little bit of independence, not a little bit ... who's thanking them? ... I mean you're every day eight hours or longer in – and getting go, go, go, go, go, go.... Somebody pushes the button ... and the line stops. It doesn't take ten seconds that the foreman is there [saying] 'what's going on?' Phew! that is, that must be a tough job.

These comments serve to underline the gulf between trades and production workers' orientations in the plant and the strong desires of production workers for more job diversity and control over their labour process. As Barney says of his operator job:

> I never dreamed that your body could get wore down the way it does from doing repetitive work, being such a big place that people always had to go to the same job every day and wear out the same joints until the joints were no good any more. When it's such a big place with so many jobs that you have to go to the same one every day, year in, year out ... wearing out your body, bending, turning , twisting or whatever the job is, the same way year in year out, it's just too hard on your body and doesn't make sense for you or the company, cost-wise, physically-wise for you, just doesn't make sense to me at all.

Clearly some views of the effects of lean production are strongly shared by all 222 assembly and trades workers, as management tries to implement more direct social control over all areas of the labour process in the plant. Tim expresses a common perception of current labour-management relations when he says:

The approach in the last three years has been drastically changed. It seems that the shop-floor management, by and large, have taken a very dictatorial approach and it seems that their hands are tied and that decisions are made ahead of anything discussed. There is no room for the union reps to have a discussion with management, to resolve problems.... It may not be really new, but it's drastically more evident that they're either unwilling or unable to resolve issues at the first step which is between a worker and a supervisor directly or within the first level of union representation in the grievance procedure. Today because of the way management works, because you didn't give them a week's notice, they seem very unable or unwilling to accommodate requests like, "my daughter's in the hospital, could you put me in for eight hours pay and take it out of my Paid Absence Allowance bank?" Today's response would be "no, you're not getting that." A few years ago, the supervisor on the floor'd say "no problem, I understand.

Formal Schooling and Job Training

Through some combination of family circumstances, the attraction of a well-paying job and perceived class bias in school, most of the auto workers we interviewed had tended to leave school earlier than their talents might have allowed. Ronnie, a machine operator, felt that his teachers disliked his lack of "proper" middle-class English and he also had to leave home to get a job. But early leaving has not dimmed his interest in creative writing:

My schooling? Grade ten basically is what I was supposed to have ... I enjoyed it when I was there basically. I was about fifteen, I ... had to leave home and started in the workforce, continued on from there to decent wages.... I enjoyed math, history, I enjoyed geography. The only one I never did get ... was English, because I just was a lousy speller and I never could get my commas and that. But I loved writing poems and stories, and whatnot. I still write poems and short stories and stuff. Haven't had anything published yet. I just kid around with it, you know ... something to do.

Sam, the lead hand, describes a retrospective sense of discrimination against working-class children which has been well-documented in research studies:

Oh, I certainly thought that I'd screwed up in school, that it was my fault, everything. Sure I could have put a little more effort into school but I don't think school is set up for those who came maybe from a blue-collar background and so forth. The system isn't set up, you know, to help or as interested in them because of their parents' experience and so forth. I'm sure the majority of, you know, blue-collar workers are, you know, my father never finished high school. It wasn't because he wasn't capable or didn't have the understanding to know that schooling was necessary. I'm sure he did, because he always, you know, gave that spiel "Oh, you need your education, it's important." But I don't think he put it in perspective. The system isn't set up, you know, maybe for working-class and lower class people.

There is a unanimous opinion amongst assembly-line workers that little formal education is actually required to do their jobs. No one holds the view that post-secondary education is needed and some see GM's attempt to hire those with greater education credentials during the early 1980s as counterproductive to tolerating the drudgery of the repeated mechanical motion of the job. As Greg, an operator with less than a Grade 10 education states:

> We have people that have college or university education. GM thought they'd be smart and hire all university students or first or second year university that dropped out. Yeah, hire them ... they're smarter people. They know how to put nuts on better than a dumb person does! I mean, their mentality is way beyond my understanding because somebody that has any ounce of brains at all in their head gets bored. Where a guy that goes to grade eight, he doesn't get bored as easy.

Harry takes the even more extreme view that no assembly jobs require substantial formal education of any kind:

> As far as education goes, the job itself, as I said, anybody can do it. The person I'm working with he has a chemical imbalance, he's still doing the job. Sometime he doesn't know who he is, where he is, and I have to work with the man and I feel sorry for him. He crows like a rooster and then sometimes he acts like a horse, he goes in up-and-down yo-yo moods, but he does the job. It does not affect doing the job. Sometimes you have to tell him whose turn [it is] to do what, but he does the job. The job itself does not require education ... it's wasted there's no way that you can use it.

Skilled trades jobs retain a lot more discretionary control even under lean production conditions, and some assembly workers have continued to aspire to them, with little success. Sam, a team leader in assembly, who is very adept at all the jobs in his department, speaks with frustration about his attempts to get into a trades apprenticeship:

> I tried, but the apprentice committee people, you know.... I have not known anybody to get an apprenticeship in there for ages. It's just, you know, I don't like that because it's just a token thing there to say, but GM have no plans of training anybody apprentice-wise. I have no idea where they are with it but many people would take advantage of it, if they could. I know that a few people are halfway through their apprenticeship and

they're back on the line ... They just pulled the plug on it, I guess? They didn't feel that they needed many. That was just a few years ago. [Interviewer: Since then has there been any ...?] Not that I know of ... some people brought in.

Similarly, Matt, a highly competent relief man, has taken the equivalent of the first year apprenticeship at GM through local college courses. As he tells it: "I applied for the apprenticeship program and then I had some interviews. I never got accepted because I had the wrong attitude."

On the other hand, those like Gerhard and Mickey, who are already in skilled trades jobs at GM, appear to share a conviction that they are more skilled and deserve better pay and benefits than assembly-line workers. This division of opinion can be compounded by lean production techniques that generally dissuade assembly workers from demonstrating any extensive knowledge of the assembly process. In fact, the likely "reward" for the worker would be the further intensification of the job and more, as well as more taxing, work. Given the routinized nature of most assembly work, a lack of apprenticeship opportunities and the prospect of having their working knowledge appropriated for labour intensification without credit, assembly workers do not generally see higher educational attainment as a path to improved jobs or promotion within GM. While assembly jobs in some auto plants have been redesigned to provide more discretionary control, GM has shown little interest in following suit. As a result, production workers interested in continuing their education are very likely to seek enrichment outside the narrow frame of reference connected with their routine line jobs.

Informal on the Job Training:
"You're pretty much on your own"

Newly hired or transferred auto plant employees are entitled to a three-day training period. As indicated earlier, this three-day term is intended by management to be used by workers as an opportunity to gain a modicum of speed in performing their job functions. Instead, many workers have taken advantage of this opportunity to learn the job of a co-worker up or down the line for "doubling up" purposes. The training conditions are specified in the collective agreement as a typical part of the old Fordist regime. The introduction of lean production methods has led to modifications of these requirements in practice. Dick recounts the current typical experience of learning new assembly jobs:

You're pretty much on your own. You get your token three days, if it's even that any more. I'm hearing less and less ... you getting a standard training time for your job. If you're switching jobs, you have to do it on your own. If you want that job now, you have to train on your own. Just go round at break and just learn the job. Earlier times they had enough men they could swing, you know, you could take your time, learn a new job. Basically, if you want to switch jobs within your own group the foremen are now telling you that you could do that, but you have to do that on your own, whereas a little while ago you could do it. I guess if you get a job posting, then the company's still contractually obliged to give you your proper training. But all the jobs aren't posted.... They won't train you now, before they would have.

Whether job training is contractually required or undertaken through workers' individual initiative, the unseen critical learning that takes place here is an informal working knowledge casually passed on from worker to worker in the pursuit of upward job mobility. All the Oshawa respondents claim that the best sources for learning about a particular job are the previous designated operator or an experienced relief man. As Matt, a relief man, summarizes, the almost universal practice within the plant is learning a job by combining the experience of others with one's own innovation:

Basically the best person to learn off of is the operator on it daily. And he would show you the shortcuts and how to do the job ... That was the best teacher for learning the job ... They'd usually show you but after a while of being on the job you'd figure out your own shortcuts. Instead of like taking a step back to the bench or something, you might carry a strap with you. Like, everybody doesn't work at the same pace or do the job exactly the same but as long as you could keep up to the line, you could pretty well set your own agenda on how to do the job.

Exchanging job-related information, the cooperative process of learning and teaching, is a daily practice among this community of GM assembly-line workers. Whether a new worker or a longstanding employee who transfers to a new assignment, under a lean production regime management relies on the informal dimensions of this training and learning process without any official recognition or compensation to employees. This is the underbelly of the "learning organization" at GM.

Life on the assembly line pulsates with the nonstop rattle and hum of heavy machinery. Despite the diminishing cracks between work-assignments, assembly workers still have some discretionary time to converse and recuperate. They use this downtime for numerous activities, not least an astonishing selection of literature, including magazines of almost any description. The majority of literature read on the line is generally related to either hobbies or other typically household- or community-based interests (see Chapter 8). Most of this informal learning provides a brief escape from the mind-numbing toils of the line.

Union-Based Education Programs

The UAW was an early leader in providing labour education programs in Canada (see Freisen, 1994; Spencer, 1994; Yates 1993) but since the inception of the CAW in 1985, these programs have been both deepened and widened substantially. In conjunction with the CAW national bargaining committee as well as through local initiatives, Local 222 has negotiated one of the most extensive union-based worker education programs in Canada. The array of programs is illustrated in Table 3.1. This includes a large array of CAW national programs of one or two weeks duration, local weekend seminar courses on a wide array of workers' rights issues, and many joint programs supported by both GM and 222. The most distinctive of these programs was probably EDGE (Education, Development, GM-CAW, Employability). This joint management-labour program allowed Oshawa workers the opportunity to take virtually any course at area community colleges, school boards and universities, with tuition and books paid for by GM. Participation in this program grew rapidly after its inception in 1993 (Burn, 1997).

While the EDGE program permitted a wide array of subject choices at local educational institutions, the CAW's internal education programs (CAW Canada web page, 1997) generally follow one of two paths. First, local union education committees design and deliver "tool-based" weekend or evening courses, covering committee person (or steward) training, grievance procedures, collective bargaining, workers' compensation, etc. Second, there are programs that seek to develop a social union cadre, including: Workplace Change and Competitiveness, Unions and Politics, Human Rights, Empowering Workers of Colour, Womens' Activism, and the Paid Educational Leave (PEL) program.

The CAW's PEL program is a four-week, adult education course first negotiated in 1977 by the UAW's Canadian Region. PEL is a residential program at the CAW's Family Education Centre, located in Port Elgin, Ontario. The curriculum includes labour history, sociology, political science and economy as well as public speaking, communications and media literacy (Gindin, 1995). PEL's goal is to build leadership within the ranks and to cultivate activists with a commitment both to the union and to social transformation. The CAW Family Education program also brings social union principles to the member's family and community (Roth, 1997).

The most recent round of CAW-GM Canadian contract negotiations, which took place in September 2002, saw the entrenchment and expansion of the CAW's current education and training programs. These now include a "Workplace Training Program" which consists of day-long courses offered during working hours, Basic Skills Upgrading, including English as a Second Language, a Retired Workers' Fund, streamlined tuition assistance for members enrolled in the CAW/McMaster University Labour Studies Certificate program, increased funding for PEL, health and safety and skilled trades programs, and an increased Tuition Assistance Program, which currently pays members enrolled in approved courses up to $3,250 in tuition and materials costs. In 1999, the CAW negotiated a program that provided financial support to members' dependent children. In the 2002 agreement the program's benefits were increased to a maximum of $1,300 per member per year and expanded to include eligible dependents of retirees (CAW-Canada/General Motors Bargaining Report: Highlights of the Tentative Agreement Between CAW-Canada and General Motors, Production and Skilled Trades, September, 2002, pp. 10-12). The original EDGE program has now disappeared, but the 2002 contract won an extension of prior monetary benefits in an improved Tuition Assistance Program, and also includes provision for 32 hours of union-based education over the three years of the contract for all members as well as a dozen full-time trainer/facilitators drawn from the workplace.

Table 3.1: Union Local, National and Joint Worker Education Programs at 222[4]

Local weekend seminars	National programs[5]	Joint programs
Steward Training	Paid Education Leave	EDGE[6]
Health and Safety	Human Rights	Pre-Retirement
Workplace Change	Health and Safety	WHMIS[7]
Legislative Agenda	Environment	
Women's Activism	Worker's Compensation	Other Programs
Grievance Procedures	Work Reorganization	Literacy (BEST)[8]
Collective Bargaining	Substance Abuse	CAW/McMaster Labour Studies
Workers' Compensation	Bargaining Issues	
Time Study	Steward Training	
Human Rights	Women's Issues	
Employment Insurance	Employment Insurance	

Sources: CAW Contact, The *Oshaworker*, McMaster University website.

Motives for learning run the gamut among members of Local 222. Of course many members pursue courses that address their job insecurity and the search for "something to fall back on." Over the past decade, employment insecurity related to the introduction of lean production manufacturing techniques has further stimulated interest in union-based education programs, both to protect their rights and to prepare for alternative employment. Computer-based courses are especially popular. As Bennie, an assembler and EDGE participant notes:

> I find anything on the computer right now, the way it's going with society, computers, you have to learn them anyways … It depends what I can learn, how quickly, what courses are available to me, that it isn't going to cost me an arm and a leg to take those courses and that as long as GM's going to offer me courses in computers, I'll take as many as they'll offer. Hopefully I'll learn as many as I can.

But the prospects of applying technical knowledge in the plant are very limited. GM claims that the use of high technology, in particular computers, is an essential skill workers must learn if they are to remain competitive in the employment market. GM has used the EDGE program primarily as an employability program which is intended as a hedge against future layoffs and an incentive to voluntarily reduce the workforce by attrition. Whether by accident or design, GM has benefitted from enrolment in computer-related courses and increased computer knowledge amongst employees. However, job-related demand for greater computer literacy appears to have remained limited. Dick, for example, does use a computer in the course of his installer's job, but as he says: "It's just a matter of keying one button. It's pretty simple, you know, there really wasn't that much to learn although it's a learning process – push a particular key on a keyboard. As a matter of fact they have a guard over the rest of the keys, so you don't press any other keys."

Other respondents engaged in union activities have ample opportunity to participate in relevant courses at the local union hall. Several times a year the Local 222 education committee, in conjunction with the CAW National Headquarters, organizes a one-day seminar for interested Local 222 members on a variety of subjects; some are tool courses, while others are issues courses that analyze the current state of labour or politics. Tim describes his experience of these seminars: "these one-day seminars, stuff at the [union] hall for committee people, for the time study was two days … the union came down and gave us sixteen hours of training on what was happening with the government changes, labour legislation and that was where I got a lot of training or more knowledge."

Our respondents also mentioned a variety of broader social interests, including greater political awareness. Anti-racist education and critiques of the excesses of capitalism are prime features of PEL and other CAW programs (Sugiman, 1994). Mickey, a white male worker in this predominantly white and male workforce, expressed his PEL experience this way: "I came back into the plant with an immense socialist vigour, especially against racism."

For production workers there is scant chance of advancement on the line and therefore little immediate advantage in acquiring advanced technical job-related knowledge through these courses. But, amongst one's peers within the workplace or union, the multiple opportunities to deepen and display one's knowledge through this vast array of union-based courses suggest that 222, if not the GM complex itself, is a "learning community." Unionized workers here find much more autonomy within their union structure and greater opportunity to exercise their intellectual muscles, in contrast to their highly regimented assembly-line jobs.

Informal Learning in Local 222

However, as adult educators know, organized courses are only the tip of the iceberg of adult learning. Like most adults, industrial workers do most of their learning informally in their everyday activities. It is workers' informal learning within their own communities that accounts for most of their continuing knowledge and skill. The CAW Education Department (CAW Canada Web Page, 1996) clearly recognizes the importance of informal learning:

> Working people learn from their everyday experiences, from their struggles for dignity and equality, and from their democratic participation in the life of the union at all levels: from local committees to Intra-Corporation Councils, to special conferences to the meetings of the union's parliament, the CAW Council. The role of the education department is to reinforce this informal education and to build on it.

Much of the informal leaning of Local 222 workers occurs at home and covers a wide gamut of general personal and social interests (see Chapter 8). As in the general population, a growing amount of this home-based learning involves computers. A number of our respondents spend time working with software packages, surfing the Internet and learning games by themselves. But, as we have previously found amongst other auto workers (see Sawchuk, 1996, 2003a), there is also a significant social learning network of computer users who share information and teach each other new programs, in spite of the fact that there has been virtually no prospect of gaining legitimate recognition for the related skills even when they are called for in the workplace. Ronnie's experience testifies:

> I've had the computer for a couple of years now, so ... done a lot. [My buddy on the line] has always helped me when it comes to computers.... The only course I've taken is through the EDGE program, so far.... I got one for the kids for Christmas. When I get time I just sit down and go through it all and see what there is and now I've got it set up so I can do my taxes, budgeting, banking, games and I can write, draw.... I once wiped out everything, I just wiped it out ... I managed to make a recovery disk and get everything back up and running, so [laughs] ... I had to borrow the disks off a guy in the plant and reinstall Windows and I had to go through all my backups on a CD and I had to make a recovery disk. It took me about a week. I learned anyways and got it all back up and running and actually never lost any programs.... I asked a few guys at work what they thought and how I could get this back and you know what I should do there and they just more or less give me a hint here and I just go home and try something else. I know a zip file is a compressed file, so I uncompressed it and "Aaah, my recovery disk!" ... I'm signing up for "Windows Intermediate" at EDGE. In the first course, basically a lot of it I already knew, it was just more or less a refresher for me, just to see if I know as much as they were giving. And I did, they just added a couple of other things that I wasn't sure, but I had it all in my books anyways.... Computer knowledge can help you, yeah, but it's more just personal. I'm just doing it for my own well-being.

But far more distinctive is the extent to which Local 222 workers commented on how they use other people's lived experience as a source of informal, transferable knowledge. Tim refers to the rich experience of a union official as a "library": "Like it's hard to believe, there's a lot of time spent ... It's almost like a library if you need something, you have to go to a library, you seek it out, without taking a course. And that's what life is generally like if you're trying to do anything, I guess, because you have to seek out who knows." There is a vast array of informal collective learning among Local 222 union activists. Darrel clearly appreciates the nature of this continual learning: "I take anything I learn in the labour movement as being educational ... period. Newspapers, past courses I've dug up for article research for as far as writing something in *The Oshaworker*, other people's collective agreements to see what we need for our locals. I mean, all of that's kind of informal, I guess, learning."

With a zeal and commitment some scholars suggest can be found only in the ranks of career professionals and executives (Reich, 1991; Senge, 1990), 222 workers' collective involvement can take them into intensive informal learning activities. Union activists like Tim can find that their involvement provides a seamless web of social learning opportunities:

[W]e can go sit down and have a beer, but we could be discussing union stuff and learning.... [W]e were supposed ... to watch a hockey game once, some buddies we brought over, [but] we never talked hockey. We talked union issues and labour problems ... right from eight to one in the morning. I'm listening; I'm learning, some of this, all coming into my mind. The whole time I did learn.... [A]fter a union hall meeting, often we go out and talk about stuff and it's work-related.

PEL's potential for stimulating continuing informal political learning is apparent when participants' observations of once-unquestioned daily experiences take on a newly critical hue. For example, Jill, our only female respondent, discusses her understanding of newspapers as shaped by her PEL experience:

I don't look at the paper in the same, same way that I used to ... when they were talking about media ... and the differences between the writers and their columns and stuff and how they ... leave things out, you know they edit it to make it a certain way so you're not really told the true story. Some people read the paper like it's gospel, now you look at it and you go 'no, that's not true,' they're contradicting themselves from like last week or the week before, or one reporter's saying something different from the other one.

A close look at the community of workers at CAW Local 222 reveals a wide array of informal collective learning activities initiated by both rank and file members and internal political caucuses. The rich set of organized educational programs available to Local 222 members provide great opportunities for members to link their extensive informal learning with courses and programs on virtually any subject of interest. There is a substantial amount of working-class learning capacity here, being applied in many diverse ways. The evidence suggests that Oshawa auto workers' involvement in organized union courses and participation in political education programs such as PEL, along with workers' continuing informal learning constitute a thriving grassroots working-class learning community. As Barney, another Local 222 PEL graduate told us:

[T]he way it [PEL] really contributes would be [reflected in] the idea of humanity that I try to show to the people that I work with. Since Scarborough [Ontario's GM van plant, part of CAW Local 303] closed [in 1993] we have a lot of [visible] minority cultures [in Oshawa] ... I try to openly show them some kind of a welcome. That's another thing from PEL too, I tended to [believe] "all trade unionists are the same." So I ended up spending some time on the [GM-Suzuki] Ingersoll picket line, I talked with those people, and I dropped into the Scarborough van plant demonstration because it was a labour strife issue and I felt ... I wanted to be there. I sort of have this sympathy ... I just sort of feel for these [laid-off] guys. They've been jerked around by a corporation that's making a lot of money and there's no reason for it. I can't see the underdog go like that.

This is not just another expression of traditional narrow trade union consciousness but rather a rank-and-file expression of a much broader identity with other exploited and oppressed groups. Political education programs like PEL have generated hundreds of labour movement activists within the plant community.

Cultural Historical Dimensions of Auto Workers' Learning

Cultural historical activity theories of learning encourage researchers to pay close attention to social contextual influences on the development of people's knowledge and skills. We all belong to communities ranging from those ascribed by attributes such as skin colour to voluntary groups of interest we can leave tomorrow. The most politically significant, modern communities have been social movements constituted by people who share common territory and economic life, as well as a common language and culture and a collective desire to transform their existing social conditions. The most notable of these are movements to create new nation-states. In our view, the organized labour movement contains some of the communities within advanced capitalist states with the greatest current potential to mobilize their members through critical learning and to transform existing social conditions in a progressive direction. Large union locals constitute communities in the sense that they are social organizations of people in immediate contact based on shared territory, economic life and language, as well as a common working-class culture (see Martin, 1995; Newman, 1993). Business-oriented union leaders and bureaucratic structures have faced justifiable criticism from progressive intellectuals (see Gaspasin and Yates, 1997) but no other social organization offers as much opportunity for learning practices sensitive to working-class needs. Labour unions like 222 are among the most democratically structured organizations. Leaders are regularly elected and subject to recall by their membership, and frequent meetings of the general membership provide real opportunities for interested workers to present alternative motions to the entire body. Labour unions generate sufficient resources from members' dues to construct and sustain organizational vehicles for popular social change, including their own educational programs.

CAW Local 222 is a community of workers engaged in many generally unrecognized worker-centred learning activities. The 222 community shares an internal economy. On any given shift, the plants throb to the rhythm of workers buying, selling and bartering a wide variety of goods and services. There is also an active exchange of labour and expertise in such areas as home and auto repair and computer or music lessons. This community has a shared vernacular language and culture (cf. Joyce, 1991) conveyed in the plant every day, in addition to regular monthly general membership, retirees' and family auxiliary meetings and publications, such as Local 222's monthly newsletter, *The Oshaworker*, a 48-page magazine distributed to almost 25,000 members and retirees. The speech of the Local 222 member is dotted with distinctive phrases: "ass-time" for downtime on the job; "gloveball" is a popular local sport; "lost time" is not a science fiction concept but money received while on sanctioned union business. In this community "brothers" and "sisters" are fellow members and not related kin. In terms of oppositional class-consciousness, a GM foreman is referred to as a "white shirt", "getting out the kneepads" alludes to those who grovel in the face of management and a "baglicker" refers to a management sycophant (see Roth, forthcoming).

Of course, there are other, less progressive community influences on the GM workforce. Continuing the paternalist tradition of founder Colonel Sam McLaughlin, the company still organizes a variety of community activities (Manley, 1986) from sports leagues to musical groups – although the GM Choir sang its last note in 2002. The competition between the union leadership and GM management for workers' communal allegiance encourages a divided class identity amongst Oshawa auto workers. However, GM management strategies, workplace reorganization and growing job insecurity have served to animate a more united culture of resistance. The local mass media have also historically had to be more sensitive to auto workers' interests than in larger communities with more fragmented labour forces.

From a CHAT perspective, the most dominant theme to arise from the Local 222 site was the wide effect of its strong union culture on both individual and collective learning relations. The local has one of the most fully developed, worker-controlled educational programs in existence. Its strong union culture has helped workers to appropriate notions of education that included them, in spite of often difficult early schooling experiences. This highly evolved workers' learning culture, dominated by a cohort of white, middle-aged males, remains one of the largest, most concentrated and well-organized working-class communities in Canada. While the opportunities for engaging in critical collective informal learning for progressive workplace and social change should not be idealized, they evidently remain very substantial.

In particular, we should note age- and skill-related patterns of learning at Local 222. While there is a general tendency for older adults and older workers to decrease their participation in organized courses, older workers at 222 appear to continue to engage in courses to a much greater extent and also to spend a relatively large amount of time in informal learning activities, as our comparative analysis in Chapter 8 will specify. We expect that this pattern is related not only to the fact that these mainly middle-aged white males who belong to a strong union have won extensive education provisions and have a relatively large amount of discretionary time beyond their jobs, but also to the encouragement of a worker community that values highly elders' knowledge and calls on it regularly.

Similarly, while skilled tradesmen often celebrate a larger measure of technical control over their jobs, they have historically played active roles in large-scale struggles for workers' rights including more educational programs for all workers (e.g., Freeman, 1981). Skilled trades' leadership in the 1996 plant occupation is indicative of the underlying solidarity between trades and production workers when faced directly with the lean production offensive. As the fall 1996 strike and a sit-in at GM Canada national headquarters in August 1998[9] demonstrate, Local 222 represents a fertile site for critical social learning and collective action to assert workers' rights.

Concluding Remarks

Creative choices persist in the immediate details of most workers' individual and collective labour processes and are chronically underappreciated by most researchers. Some prior workplace ethnographies have managed to reflect the dynamic nature of this continual modification of specified work tasks (e.g., Kusterer, 1982). Even assembly-line workers facing detailed tasks formally designed by engineers must pace themselves, consider alternative routes to completing tasks and frequently deal with deficiencies in the design of parts or engineering. Local 222 workers continue to have daily options to initiate more time-saving, work-to-rule or destructive responses to their designated jobs. As seen in these interviews and our participant observations, assembly-line workers' knowledge gained through informal peer training, accumulated through trial-and-error and information acquired through a "dialogue" with the problems posed by the assigned tasks are all forms of ongoing intellectual work

However, the onset of lean production with intensification of technical tasks and a more authoritarian management approach has probably diminished line workers' discretionary control. Lean production has undermined the provision of formal training for new jobs and increased reliance on individualized informal methods of picking up new job knowledge "on the side," in ways comparable to the atomized garment workers we will meet in Chapter 7. GM's cost-saving strategy of downsizing by not replacing full-time retiring Local 222 members and increasingly relying on a contingent army of temporary part-time workers without effective bargaining rights is also closely related to company efforts to reduce the general education provisions established under the Fordist regime. In particular, PEL and the terminated EDGE program, through which numerous forms of education not directly related to current job tasks are eligible for company funding, must be seen as extraneous by a management increasingly fixated on just-in-time delivery systems.

In this context, the company and many workers are likely to opt increasingly for individual instrumentalist responses to learning. Computer training courses at local colleges, for example, do have some limited prospects for application in the plant but are more likely to attract workers looking for preparation for alternative employment outside. Extensive forms of individual informal learning for fulfilment beyond the job, such as the elaborate horse racing system discussed in Chapter 8, may also be proliferating.

But collectively based education programs and union-oriented informal learning practices still thrive and provide a stronger foundation for progressive struggle for change than at any of the other work sites we have studied. The CAW/GM plant is easily the richest of all work sites in this study in terms of the provision of union-based education programs. Oshawa's auto workers have choices well beyond those of their counterparts at most unionized workplaces across Canada. Courses remain readily available even in those areas apparently unrelated to work activities and local leaders have fought to retain these provisions. The combination of PEL and the local's weekend and evening labour courses, continue to provide an exceptional platform for collective worker education and learning.

Whether these rich organized opportunities for collective education and worker solidarity will continue to sustain comparable *informal* learning opportunities is the unanswered question. Our research cannot provide any definitive answer. But Local 222's workers clearly have the financial ability, the time and spaces and the easy availability of relevant learning materials to pursue an abundant informal labour-oriented learning agenda, if they so wish. In myriad spheres of their lives both in and outside the workplace, these are avid working-class learners. Local 222 members pursue learning in all terrains that capture their fancy – from repairing their automobiles or renovating their homes to learning about a particular point of interest in history, mathematics, studying an aspect of animal biology, delving into computers or languages. Our interviews certainly revealed that informal learning was often viewed by participants as an individual escape from the rigours and monotony of assembly-line work and as an outlet for job-related frustration and insecurities. But many PEL-trained activists and others at 222 deeply understand that engaged dialogue with other rank-and-file workers remains imperative to sustain any real democratic cultural, educational and political change through this community. As Pete, the young assembly-line worker profiled at the outset of this chapter, puts it:

> You've got to understand. If there's a [social union] agenda that you want to put across to people and you want to build a movement, you have to have some knowledge of what the issues are. You have to be able to explain them on the shop floor if you want the support because as a union we're only going to move forward or push an agenda if we have the support of the people.... We're going to have to pave the road before they'll drive down it. It's a dirt road and they're not ready to drive down some of these roads yet, but they will, or we're going to lose. We're at a bit of a crossroads. We have to change. We can't stay the same, but we can't forget our past and we can't give up the gains we've made. So this means new ways of fighting for things and sometimes the new ways of fighting things are going back and doing the old way. For example, young guys who have never been on strike have to learn how to picket and how prior strikes contributed to increased workers' and civil rights.

Notes

1. Big Three plants in St. Catharines and Windsor Ontario have lost significant numbers of employees. Quebec's only auto plant in St. Therese is slated to permanently close in 2006.

2. In the mid-1990s, after a series of electoral defeats that saw the Autoworkers caucus membership plunge dramatically, the remaining handful joined the Democrats caucus and effectively ended almost half a century of this particular rivalry.

3. See Traill video (1997). This is a home-videotape made by a Local 222 GM member who assigned

himself the task of documenting the 1996 strike. Hundreds of workers purchased copies of this tape as a memento of the strike when they returned to their jobs.

4. Note there are additional labour-education courses available at the federal level through the Canadian Labour Congress (CLC), at provincial labour bodies across Canada and through local regional labour councils (see Taylor, 2001 chapter 6). Local 222's course offerings have continued to expand over the past few years, through McMaster University's Labour Studies program, which offers a CAW/Mac Labour Studies Certificate. Current course offerings include: 'Labour and the Internet,' Globalization, Media Studies and Terrorism.

5. With the exception of Paid Education Leave all national CAW programs are one to two weeks in duration.

6. "EDGE" is an acronym for "Education, Development, GM/CAW Employability." EDGE was initially established as a Oshawa-based, union-company pilot program aimed at enhancing employees' skills. Although it was discontinued in 2000 its key benefits were extended in the Tuition Assistance Program, which is governed by the GM-CAW Master Agreement.

7. WHMIS (Workplace Hazardous Materials Information System) is a federally-mandated workplace hazardous materials and safety awareness program.

8. Basic Education for Skills Training was the largest literacy program in Canada. Coordinated by the Ontario Federation of Labour until the late 1990s, BEST collapsed due to the withdrawal of all provincial government funding. The National office of the CAW has since assumed responsibility for a revived, if modest, BEST program.

9. See "GM workers sit until they get meeting" (*Oshawa This Week*, 9 August, 1998). This article describes a sit-in at GM Canada's national headquarters by thirty Local 222 shop-floor representatives who were frustrated at their failure to schedule a union-management meeting.

CHAPTER 4

Building a Workers' Learning Culture in the Chemical Industry

John Carsons is a batch processor in the high performance department of the chemical factory. At 49, he has lived a full life and, as a white, middle-aged male with a host of working experiences to draw on, he represents the core workforce of the chemical factory in many ways. He was born into a strong, close-knit Irish-Canadian community in the Canadian Maritimes but poor employment prospects on the coast caused his father to move the family to the bustling city of Toronto in the 1950s. There they settled into a small tenement in a working-class district; a neighbourhood full of Martimers as well as recent Italian- and Portuguese-Canadians. His father was a former military man and a stern disciplinarian. He was also a heavy drinker, all of which weighed heavily on the family. To escape, John spent a lot of time running the streets, and as we talked about his high school days he recounted the many friends who ended up in trouble with the law. By age seventeen John recalls, one of his most important discoveries was simply that he had to find a way to escape poverty and the hard life: "I said there's no way I'm going to be living like this. I'm going to turn my life around, but I've got to do the turning around." Indeed, his father's drinking contributed to his sense of the need to take control of his own life and will a new one: "The one thing you have that makes you equal to everybody else is your will-power. If you don't have your will power you got nothing." Indeed, this notion represented an anthem for many of the men at the chemical factory and particularly those who became active in the union local: alone or collectively, assert your will, create change, make things better.

John's youth also shaped his view of his own learning capacities and ability to shape his surroundings and future. He successfully completed high school with a genuine fondness for chemistry, although he tired easily of authoritative teachers. He recalls doing well in school, and early on made plans to enrol in the local university but, in part due to his father's declining health, John was forced to abandon this plan to find work to support the family.

Sending home paycheques regularly, John worked and travelled the length of Canada working in a variety of jobs through his twenties: as a miner in Thompson (Manitoba), as a labourer in Vancouver and Calgary, and finally as a scullion and wheelman on the giant Lakers in Ontario. It was on the lakes that John had his first taste of trade unionism, entering into and learning avidly about the history and operations of the Seafarers International Union (SIU). He liked the way the union made employers think twice in dealing with workers and it allowed him to learn the way he liked best: by working side-by-side with experienced seamen who knew what they were doing:

> *You have to learn things by doing it, you can have the book beside you, but you have to get your tools and do it ... I did good in school settings but I don't retain it.*

By the 1970s, John had decided to leave the boats, and he took several different jobs in a growing chemical industry desperate for workers. Now he applies his learning abilities and his general confidence to work situations. He is skeptical about the extent to which the company's rhetoric of self-directed workforce matches reality, although he has attended an enormous number of internal and external courses. Across the workplace, as with John, there appears to be a critical enthusiasm for learning opportunities, as well as a confidence to affect positive change in production processes. By the end of our interview sessions together, there is a range of clear themes that had emerged. For John, even the most high-tech workplaces don't challenge workers enough. His own experiences tell him he is "trouble" when he's not challenged, when he feels his skills, knowledge and experience are not being used. He sees the same thing in the faces of his co-workers on the shop floor as well: "They're bored with their jobs. Bright, educated guys just want to say 'screw it' because a lot of times the work is below them."

Introduction

Although much is made of how management-based initiatives such as Quality production, Organizational Learning and ISO certification affect new "knowledge-based", "worker-driven teams" and so on, these types of claims are more often asserted rather than critically examined. They are only rarely examined for their effect on the actual learning and work lives of employees from their own perspective. Understanding the economic context is part of this critical examination. Situating workers' work and learning at the chemical factory is essential for appreciating the organizational forms and political-economic roots of these workers' practices.

According to Statistics Canada, the domestic chemical sector as a whole is riding a long wave of increasing domestic demand and stable market pricing which, after a slight dip in the first half of the 1990s has produced more or less stable employment levels and renewed capital investment (Industry Canada, 1997a, see Table 4.1). Indeed, Industry Canada is forecasting continued growth in each of these two areas for the near future.

Table 4.1: Canadian Chemical Industry Trends

Year	Employment ('000)	GDP ($M x '000)	GDP/Employee ($ x '000)	Capital Invest. ($M x '000)
1988	93.9	7.3	77.8	—
1989	95.4	7.6	79.3	—
1990	94.9	7.6	79.9	—
1991	91.5	6.9	75.4	2.1
1992	91.1	7.2	79.1	1.7
1993	87.5	7.5	86.1	1.7
1994	84.3	7.9	93.6	1.3
1995	83.8	8.1	96.5	1.1
1996	84.6	8.2	96.5	1.4
1997*	84.8	8.2	97.0	1.9

Source: Statistics Canada, SIC:371 "Canadian Chemical Industry Statistics Handbook," 1997 <http://strategis.ic.gc.ca> (Industry Canada)

The chemical company at this research site supplies other industrial sectors and is part of a global conglomerate. Its Canadian division consisted of several production plants employing just over 2,000 workers. This work site is easily the most economically stable of any examined in the WCLS project. The broader corporation is consistently in the Top 200 net sales and describes itself as having balanced domestic and globalized production including factories throughout North and Central America, Europe, South-East Asia and China. As a whole it has a very strong record of growth. Within the North American context of this company, the division to which this plant belongs was particularly strong, buoying weaker areas of the corporation as a whole with net sales, gross profit and net income setting the pace. The basic statistical profile begins to describe in quantitative terms the economic context within which work and learning take place. However, policies, managerial practices and the overall "culture" of the company and firm also play a very important role. Indeed, the company typically makes public its general operating procedures around its desired "workplace culture" and specifically orients its policy to creating the much celebrated "learning organization":

> We strive to be a continuous learning organization, which is constantly adapting, expanding and creating. By focusing on competencies, all [company] associates can develop a clear understanding of the requirements for success, and individuals are empowered to take responsibility for their own professional growth.... On the job experiences, rather than classroom training, are emphasized. This establishes the relationship of work performance to promotability. (Company Employee Development statement, 1998: 4)

In particular, the company's commitment is said to run across each of its business divisions in an attempt to create more elaborate "career ladders" for most, but not all, employees.

> [The company] has committed to employees that all U.S. exempt position vacancies (those typically requiring a college degree or equivalent experience) up to a middle management job level will be posted weekly ... By posting jobs weekly, [management] helps employees have greater involvement in their own career planning. (Company Job Posting statement, 1998: 5).

The presumption of this policy statement is, of course, that "career laddering," despite its rhetorical commitment to "all associates," does not extend to production employees; an issue that workers in our research confirmed.

However, this does not mean that the company does not seek to incorporate production worker's knowledge as part of its ongoing organizational policy. Beyond mandated union representation on the company's Steering Committee, the company has supported efforts to involve workers in work design and redesign in keeping with its policy on "Employee Development" and "Empowerment": "[e]mpowerment allows those people who work in the position closest to where problems exist to have the ability to take corrective action." (Company Empowerment Statement, 1998: 4). Over a decade previous to our research this philosophy of worker involvement centred around the company's attempt to set up a leading-edge, high performance, environmentally friendly section of the plant where experiments in "worker-led" production could be made. Now, with what one union executive member described as "gentle prodding," a similar arrangement has been made to have workers play a significant leadership role in the design of another new facility currently being built on a land site adjacent to the existing plant.

Labour processes, as in many workplaces, vary from department to department, but there are no conventional continuous run assembly lines. Most work is organized as batch processing with the work of individuals and small groups linked together with material handling work. Exceptions include, again as with most workplaces, maintenance work, which is organized around a mix of planned and unplanned activities in which trades people largely schedule their workdays themselves and move about the plant independently with little direct supervision. The other exception is the newest department, originally planned with the input of a select group of workers who then became something approximating what is known in the literature as "high-performance" production with worker-driven and team structures. The reality was that it was organized around the batch processing system where there was less direct supervision and workers had some say in new hires to the department, dealings with maintenance and basic technical changes. In the introduction to this book, we drew on the work of Lazonick (1991) and described the managerial strategy as "innovative" with employment on a slow but steady rise, as well as a rating of "high" in terms of training policy and relative wage levels. Clearly the labour processes, strategies and policies are among the most progressive that we examine in the book and this cannot be separated from the economic factors identified above. Although at the same time, our task in the analysis will be to sort out the shop-floor "facts" from the many possible "fictions" contained in managerial policies and public statements.

In terms of a managerial rhetoric of employee development, programs and opportunities this site has some important similarities to the college site in Chapter 5. Likewise, there are some important similarities between the make-up and strong union culture in this site and the auto assembly site discussed in Chapter 3. Both workforces are predominantly white, male and middle aged; an independent and vocal group of people who know their legal rights in the workplace and vigorously defend them collectively. However, on a broader scale it emerges that the work and learning, as well as struggles of the local union overall, occur within a unique intersection of economic and sectoral factors, managerial strategy as well as (union and organizational) cultural factors.

Profile of Workforce and Interviewees

We interviewed twenty of the 144 production employees at the plant (14 percent) (see Appendices). Previous to our research the plant had experienced modest but steady growth in production employment levels and, just prior to our arrival, added the high performance department and began the construction of a new warehousing facility. As in the other sites we were careful to construct a representative sample that addressed key social variables. The average age in the plant was forty, and our sample paralleled this (average age of interviewees was forty-one) with a range of twenty-five to fifty-eight years old. We were somewhat more interested in the experiences of newer workers (which is not to say *younger* workers as they had on average substantial previous work experience) because of the firm's claim to be a "learning organization" that provides extensive training. Approximately half the interviewees therefore had under ten years' seniority. Of course, we sampled across all major departments with attention to both newer and older production areas. We were also careful to select a mix of those who were active and non-active in union work (eight considered themselves to be "active", twelve "non-active"). And finally, it is important to note that at the time of our interviews only three production workers were women; and we met with two of them.

A Profile of Learning Amongst Chemical Workers

[The Company] recognizes that in competitive global markets, employees make the difference between an ordinary company and an excellent one. Employees come to [the company] from different backgrounds, experiences, countries, cultures, races and genders and have different styles, talents, attitudes, values and skills. These differences together can result in greater creativity, better insights and improved decision-making, all vital to success in a competitive marketplace. By working together in a climate built on the value of a diverse workforce, the understanding and valuing of differences, open communication, teamwork, trust, personal development and recognition, we can realize our full potential – not only as individuals, but as a company. (Company Diversity Statement, 1998:6)

In our research, it was the chemical site management that spoke most strongly about an expansive view of learning and knowledge in the workplace. Our inquiry focused on how – and indeed if – these stated views, policies and programs were realized.

Formal Learning

There was an enormous variety of formal schooling credentials in the plant, although as seniority levels dropped the levels of formal schooling

rose, suggesting that, in the last decade, a clear pattern of credential inflation had emerged at the company. Many senior workers, typically those in
their late 40s and up, had not completed high school (in some cases had
not even started it), and in interviews they talked at length about why. The
most common reason they gave for leaving school was simply that they
felt further schooling was unnecessary for a factory work-life: a well-paid
and secure job did not require a high school diploma and would not affect
their ability to contribute. Indeed, these older workers still felt this way
despite the influx of highly educated workers in their 20s and 30s who
were now coming onto the shop floor. Other reasons for not continuing
with more formal schooling included the need, as we saw in our opening
vignette of John Carsons, to support one's family or help out with household income where the breadwinner had left or become ill or injured.
Moreover, both the men and women with whom we spoke, said they simply disliked school, it didn't teach them what they needed to know or even
how to learn. Mostly, they preferred to learn with others, learning-in-do-
ing, hands-on. Several interviewees were therefore able to select paid
work early in life, or, in several cases, to follow their passions in music,
mechanics, travel, running a small business and so forth, after which time
they needed to earn regular income but felt less suited than ever to a return
to school. In contrast, many newer and younger workers had completed
high school and, in many cases, post-secondary schooling and even professional programs. The formal schooling credentials among workers interviewed ranged from grade six to trades apprenticeships, and professional and graduate degrees. Among those with post-secondary schooling,
the areas of study were broad: computer engineering, mechanical engineering, education, law enforcement, business administration, accounting,
political science and psychology.

More specifically, regarding interviewees' perspectives on educational
credentials and their relevance for work, workers made comments such as
the following: "People have too much education. We've got post-secondary education and skilled trades that are doing general labour jobs" (Dan);
or "[The company is] hiring people mostly with university and college experience, teachers, accountants … and these people know their rights
too!" (John).

These statements introduce a number of important points that we will
examine in this chapter. Hiring at the plant now depends increasingly on
education attainment and a high school diploma is now a minimum hiring
requirement. Previously there was no minimum requirement. In terms of
formal learning, clearly, there was a wide range of rich experiences and

abilities on the shop floor but exactly how these credentials affected and/ or were applied to production was not straight forward.

Workers with whom we spoke differed on what they thought of formal schooling, a few thought it was an intrinsically valuable and highly practical experience while others saw it as a waste of time. Interestingly, these views were not distributed systematically across those with higher or lower formal education either. Indeed, as often happens, workers sometimes held two or more conflicting perspectives. The different views on the usefulness of formal schooling can be summed up in four basic ways.

First, many interviewees felt that formal schooling up to grade 10 was useful for the "general skills": reading, writing and math, as well as helping you to "know your rights," that is, feeling entitled to a decent work environment. They also felt that formal schooling was very useful in terms of understanding health and safety issues. Factories in general can be dangerous places, but working in a chemical factory introduces many additional health concerns. The workers who had formal schooling experience in areas related to health and safety (e.g., technical certificates, apprenticeships, high-school chemistry or technical courses) often made comments like, "You really need to know about the chemicals for health and safety ... literally if you don't you could blow the whole thing" (Sashi); or "My experience in high school chemistry has helped a lot here ... as did on-the-job experience in similar work" (Lee). Third, workers all felt that formal schooling credentials were useful for getting the job in a competitive job market. For these workers, "Education looks like the big thing now with a grade 12 minimum ... If they had the testing in place that they have now a lot of these guys [who arrived before applicant testing; the grade 12 minimum requirement] wouldn't be here" (Sean); and, "People are generally smarter than the jobs require but still, with computers, the way things are going it seems like the more education you have the better ... you know you might get laid off and you're out there looking" (Teddy).

Finally, many interviewees felt that formal schooling was simply useless. What was needed was job-based experience and possibly a select set of courses (again, health and safety was key along with computer training). Comments included: "I don't have any [education]. I don't believe in it" (Tom); "Grade 12 is not really needed. I don't think some of the guys have grade three ... you have to be able to read, write ... you know, maybe math, but it's just basic [as far as formal schooling is concerned]" (Raymond); "Yeah, [I took] college and university courses, but grade 10 gave more of a stepping stone to the real world than the others ... you sit around a lot at college and university" (Dan); "School does nothing to help you work in a blue-collar job" (Dean).

The latter perspective was most common, indicating that in many ways formal credentials do not particularly aid in the actual work. At the same time, all workers, no matter their formal schooling background, felt that other ways of learning and other types of practical skill and knowledge were invaluable.

Company Training

[The company] don't hire part-time, as they spend funds on training, because people are skilled they can't just replace them. It is too expensive salvaging batches. If you make one mistake on a 1,600 gallon batch, it costs a lot of money. (Sean)

It's amazing the training they have! (Sam)

Oh, I don't know ... there are lots of different courses that you can take. Some of them are mandatory and some of them aren't. (Alan)

[We] place considerable emphasis on employee training and development. Each full-time, temporary, part-time, job-share associate[1] is involved in a training process, including associates paid by [the company] on a contract basis or through an employment agency. (Company Newsletter, 1995: 8)

I fell asleep. (Tom)

This company appears to run an extensive training program: a company training schedule at the time of our research listed a total of 436 different courses. However, by listening carefully to workers we learn that these high numbers do not tell the whole "training story" at the plant. This is because the work site actually serves as the national headquarters for the broader corporation, housing within it a great number of executive and (non-unionized) professional administrative and clerical staff. Thus, while the unionized, production workforce makes up approximately one third of the total workforce at the location (about 37 percent), less than one fifth of the training courses are accessible to them as production employees. More importantly, however, of the courses meant for unionized workers, between 70 percent and 90 percent were what was called "OJT" (on-the-job training) courses. Although this is extremely valuable training according to workers, it is not course-based, and subsequently receives neither the same respect nor the same resources. From the company's perspective, OJT training helps the firm present itself as "leading edge". OJT training also qualifies as a legitimate training item under the International Organization of Standardization (ISO) and Quality Standards[2] (QS) guidelines which also bolsters the company's image while costing little. The focus for ISO and QS certification does not, in fact, concern the quality of training at all, but with the documentation and control of training. As one worker said, "They document. We train." (Tom) and thus, although both workers and management share a general interest in learning/training, their interests and actual roles are quite different.

One way of comparing the differences in training between unionized and salaried employees would be to compare the resources spent on training. Most training for the unionized workers, predominantly OJT courses, are supported only through the worker's "training rate": at the time of the research, 25 cents an hour, on top of regular pay, after the first two hours. This would provide, for example, four days of personal, one-to-one, hands-on training for a little over $6 minus a marginal drop in the output of the worker doing the training (however OJT trainers were still required to achieve their basic production quotas): an incredible bargain when compared to corporate training of any kind. Another way of looking at it might be to ask what kind of cost would be involved were an employee to try to learn without the OJT training provided by fellow workers? Again, the actual costs of ruined batches, injuries and accidents, and of course the accompanying loss in sales as production costs rise, all help to reveal the bargain offered by the unionized workers who double as trainers.

Nonetheless, interviewees maintain that there are courses available and that the company seems to think training for workers is important. In an article about the company's Training Development and Education department (1993), they described amongst other things how the OJT approach is an "effective, inexpensive developmental alternative" to courses, so here again we can plot some diverging interests and goals in regards to training rhetoric. Overall, cost savings would seem to be important motives for the company's training structure, while the company's preparation for the ISO[3] and QS certification processes[4], with its emphasis on formal documentation[5] serve as an additional source of motivation, none of which has anything to do with effectiveness of training, expansion of developmental opportunities for workers and so on. Workers interviewed were quick to add that training in the mid-1990s emerged in parallel with the focus on controlling health and safety costs and the company has, in fact, made several public statements regarding the nature of this commitment to improving its health and safety bottom-line.

> Our objective is to prevent accidents by ensuring that individuals recognize and avoid unsafe practices. The concept is being embraced enthusiastically throughout [the company], and we are rapidly expanding this approach globally. Behaviour modification appears also to have great promise for addressing occupational health and environmental safety issues, and we are exploring ways to effectively extend the concept in this regard. (Company Ergonomic and Behavioural Leadership Statement, 1998)

110

Further to health and safety training focusing on labelling, first aid response, hazardous materials handling etc., the training includes general skills training and orientation in, for example; computer operation, charting, quality, and so forth. The workers interviewed generally felt that the health and safety courses and the monthly safety awareness meetings were the most useful of the company-initiated OJT learning scheme although some workers have criticized the safety awareness meetings for the lack of action taken on the issues that arise. In addition, most workers preferred to have a departmental health and safety representative because members felt departmental representatives were more accessible and that they would have a better feel for the problems particular to their own departments. It was felt that the local's full-time "health and safety" position needed time in which to develop the position and routines to be effective and to work smoothly with departmental representatives.

One of the most basic principles that emerged from the interview data around workers' participation in company courses and feelings of their general usefulness is the notion of "use it or lose it." Almost all workers talked about the futility of trying to learn something and then not being able to actively use/develop the skills on the job.

> Training is wasted on people who don't work with computers either. You use it or lose it ... Classroom training was good but they should have kept it up. If you don't work with it or study it on an ongoing basis you lose it. (Valerie)

> And there are SPC [Statistical Process Control] courses and all that you know. You learn how to do the charts and different things like that. But I don't get into that too much even though I want to because our group leader does our charts anyway, so it's not something that you really have to dwell on. If you're not going to spend a lot of time doing it you lose it anyway. (Teddy)

> Computers are a tool. I use them, but I don't dive into them ... I grew up with a sliderule, not a computer ... if I need to learn something [about computers] I learn. It's time consuming and I have better things to do with my time. (John)

The workers interviewed described how they would quickly lose their motivation for learning when they realized that they would not be using and developing what they learn, and so some see the courses as rather a waste of time. According to those interviewed, training/education programs should never educate for the sake of education alone. They must be seen to have a practical use, or they do not inspire.

Again, we see a general relationship between the ability for people to integrate their learning into their job and the authority they have in the workplace: "Training isn't really available to everybody. It depends on their position" (Sam). Thus, closely related to the idea of "use it or lose it" are the opportunities to use knowledge, skills, learning on the job. As these workers say, while some aspects of the job require judgement and skill, other aspects are "idiot-proof" (Pete). Interviewees often mentioned that where workers are allowed to use their skills and knowledge most, they are often most interested, involved and effective. That is, "people just do a better job if they are more involved" (Teddy), but the way the job is set up sometimes does not allow workers to use their own judgement or training. In these cases the workplace might offer opportunities to learn but less opportunity to use or practice what was learned. "Always found whatever [the company] tries to teach us is informative and good, but they're bragging we're a self directed work force, which is bullshit. I was hired as an employee, not an associate" (Sam).

These workers describe how quality training is really not very applicable to the job partly because workers are still not allowed to use their own discretion at work:

> When we first started with all this quality stuff, I agreed we need quality for customers. We had quality courses, everyone in the whole plant. I find it hard to describe how they were applying it with us ... It's hard to describe to someone who hasn't been there. Sometimes their standards are off the wall. (Raymond)

> In general people are not allowed to use their knowledge. The work is simple, maybe too simple. (John)

Varying degrees of learning and use of judgement are required on all the jobs but, as workers said, the amount of learning/judgement that people can use regularly on the job is also related to the department in which they work. Trades and maintenance[6] and the high performance[7] departments appeared to provide the greatest opportunity for the use of skills, learning and judgement. The idea that the work in some departments seems to allow greater use of skill, experience and judgement is also associated with seniority and higher job classifications, since often those with high seniority or specialized credentials get into these particular departments in the first place (see below).

The other main trend that workers discussed concerned the idea that normal day-to-day production and the design of production was, as Pete comments above, "idiot-proofed," while the unforeseen, troubleshooting, preventative actions that workers had to undertake (regularly) did require considerable judgement, knowledge, experience and creativity. The trend then (as illustrated above) is that where the company can make decisions about how things are done – and when things run smoothly – the work is simple or, as John would say, "too simple." Where engineering and management ideas ran into problems (large and small), the development and use of workers' collective forms of informal skill and knowledge kept things running.

As there is no evidence here to suggest that workers with higher formal education do their learning any better or worse than other workers, this puts into question the thinking behind the inflation of schooling requirements for new employees. It might also explain why the industrial work does sometimes become more complicated, despite the companies' best efforts. When complicated machines (designed to make work idiot-proof) fail to deliver, it takes more "smarts", learning and expertise to get them going and keep them going.

Interestingly, many workers commented that one of the most important aspects of the company courses was not so much for their work at the chemical plant, but for potential future work, should the plant close down or lay-off.

> I take [company] courses to broaden your horizons. You can never have enough courses – the job market is so tough out there. (Pete)

> [T]he company will give you computer training but won't pay your hourly wage unless you need it for your job ... You know this company could be gone one day and you know I have the computer to fall back on, I have the [health and safety courses] to fall back on, you know. (Sam)

The appeal of the company's courses for the purposes of job-hunting is most strongly expressed by the younger or less senior workers with experience in the competitive job markets of the 1990s. As John, the worker in our opening vignette, notes though, he too can see the effect that the economy has on workers and their work. "Workers seem more mature now. They appreciate work. The economy has something to do with that" (John).

Many important observations and practical suggestions about company training/courses came out of the interviews. Most focused on who was doing the training and how it was done. With very few exceptions, those interviewed said that attempts at worker-driven production should also extend to training, meaning more workers involved in the planning and presentation of learning programs was best. "Sometimes workers are sent on courses to learn the machine and then bring it back to the workers. They used to send office people, but now they send more workers who know what is needed and how it's needed. They need more of this." (Pete). This involvement was said to eliminate the tensions that can arise when outside instructors have to talk about the actual production process on the shop floor – that is, the workers who know this subject best. "If it's company-oriented it tends to be bull. They don't practice what they preach. The union courses tend to be more real world. They're [the company] living in a perfect world that doesn't exist." (Dan). Companies' cultural training (e.g., quality workplace, conflict resolution, communication) were usually judged harshly by the workers interviewed, principally because it was not applicable for production. In contrast, job skills training was judged more positively, particularly when workers had the freedom to apply their new-found skills. This emphasizes the previous comments that making things "hands-on" was important because it is a part of rendering knowledge relevant.

> There was a great deal of training, a lot of it wasn't hands-on training unfortunately, we went for a lot of team building seminars, quality processing, SPC (statistics) and all this kind of stuff ... we took one, I think it was a team building ... it was conflict resolution. That sort of thing. It was interesting, but most of the training for the most part was just a joke. (Dan)

Some workers also mentioned some basic barriers to courses that involved scheduling. Here one worker describes trying to get into a course. "I was on midnight shift before, so I was trying to switch a week off. The supervisor wouldn't allow the switch for some reason. I said 'you're the silly "you-know-what" who wants me to take the course and you won't give me the time to take it,' so three months later I'm taking it" (Jane).

Giving supervisors control over who gets training was therefore also seen as a major barrier by most workers. In addition, there was the idea that training opportunities were not evenly distributed according to company policy.

> [The company] vowed forty hours of training for everybody a year, but you don't see that no more. They are going back on this a bit. Some training is open but not paid. It worked out to forty hours in my first year, ever since that though, I haven't seen a course. (Sam)

> Yeah, well I think the reason for that [cutting back on paid computer training] is people were taking it, well, I know some people are taking whatever they can get, and I know this one guy – it's got nothing whatsoever to do with his job and he was getting paid to do it so I think they [the company] were thinking he was just abusing it. (Dan)

The idea of taking company courses in preparation for finding another job resurfaces; this time in connection with shrinking opportunity for paid company courses.

This point develops the idea that the work in certain departments and positions seems to allow the greater opportunity to learn, use knowledge and gain/use experience. In general terms, many workers with either special, production-related credentials (e.g., in the maintenance department) or generally higher seniority (e.g., in the high performance department) do work that is designed (by the company) to require more skills and thus more paid training (by the company). This means that many other workers (whether they prefer it or not) are locked out of much of the paid training (although training is still available on their own time). It means that those with the seniority and experience have more opportunity to learn and further develop their knowledge and experience.

This brings up the idea in which many workers expressed a direct interest: internal apprenticeships. Many workers said they would like the opportunity to become a more knowledgeable, skilled (and adaptable) worker by learning other jobs in the plant. Several of them quickly added that this extra skill and adaptability should be compensated with higher pay. In the final analysis, the collective skills of the workforce would be greatly enhanced, well worth the raise in pay, however, the adaptability or flexibility could erode traditional jobs, skill and expertise among individuals which in some ways makes the membership more interchangeable, increasing interworker competition and possibly dividing the membership.

Informal Learning on the Job

According to the interviewees, the most significant type of learning in the plant was informal: learning in a community of others, by oneself, from a supervisor or lead-hand, asking questions: what many describe simply as "getting your hands on." This was learning about the actual world of production, complete with the complexity and contingencies of technology and collective actions – exactly the type of knowledge and learning lacking in many courses. Alan, a lead-hand in the shipping department describes it: "You learn as you go. There is no course. My department is a little harder to learn. There's a lot of computer stuff. It is not hard to work the system computers, but it takes a month or so to pick it up … but you have to keep your head up" (Alan). Almost everyone interviewed talked about learning from co-workers as the best way to learn, especially when first starting the job.

> The lead-hand really helped me a lot, but pretty well anybody on the floor would help you out. They start you out with small, easy batches. Now I could do any batch. (Jane)

> They just had people who normally did the job and they spent time with you, showing you what to do, pertaining to how the ticket is done, how

you put the stuff in. Your weights and everything is already written on a card; you just have somebody showing you. Somebody by your side all the time showing you. (Teddy)

I share computers with other people whenever, wherever its available. I have booklets that explain all the programs in detail and that is okay if you have time, but I need more time to be able to work on the computers. Trying to find room at a computer terminal is a problem. (Lee)

Informal learning was important but should not be considered a panacea, largely due to the way that it was affected by basic material constraints. As the workers quoted above suggest, time and space – and we should clarify this by saying the organizational rules, divisions of labour and authority that structure their distribution – can be an important barrier to informal learning. Courses, for computers for example, were thought to be of some help (but not in isolation from application or the chance to discuss with others) because classes helped to get people started and give them some basic access. Courses and "safety awareness meetings" were useful largely because trial-and-error learning can be dangerous in some cases. In most other topic areas, however, workers learned best on their own or from other workers when appropriate conditions of freedom, open discussion and practice could be established (without penalty by management).

Several more experienced workers also commented that this informal method seemed best, especially for newer workers. "I think it's better this way because the union employees can show the person, and the person can see the job actually happening and they can ask questions" (John). It permitted a view of the broad system of production activity *in the genuine context*. It allowed one to generate meaningful questions about the nature of the work system which, in turn, led to changes in the workplace and, in some cases, to an emerging consciousness of one's class position within work and society. In other words, the informal learning taught not only skills, but the general culture of work on the shop floor including how people are to act, how people can and should be treated, and so forth. Thus, for all workers, informal learning was part of a process through which people expanded their social network of contacts and participative opportunities, but informal learning also taught them about conflicting class relations and their own broader standpoint in society, each of which have important implications for the production process, workplace change *and even* effective trade unionism. In term of work and learning relations in the production process specifically, however, though billed in company literature and vision statements as a "self-directed workforce" generally, this is, according to workers, not completely accurate. Extending the self-directed workforce principle beyond the high performance department would have a positive effect on worker learning and skill development, though interviewees had some apprehension around this. According to them people are often more comfortable with the "devil they know". In some cases, workers have actively chosen to work in departments that are not "self-directed" and "high performance" because they see this as a facade. In the end, however, greater worker involvement and support in all aspects of training provides more choices/control of information and knowledge for the membership, and, in this sense, can be considered a more "efficient" and effective form of learning and work.

Union-Based Learning

Almost all the workers interviewed learned through the union (on courses or informally). A strong, positive union learning culture within the workplace was linked closely with a series of progressive executive leadership influences and supported by a stable employment situation linked with relatively progressive company learning policies. Indeed, having sampled workers with a range of levels of union involvement, it seemed clear that those who took an interest in union issues and union courses also took significant interest in other types of learning (informal, company courses) and topics (see also Sawchuk, 2003b). People learned about workers' issues through the union at the union meetings. They obtained materials to study alone, took union courses, became involved in union campaigns or issues, and spoke to other members about union issues within the normal routine of the day. This last method of learning was spoken of with positive enthusiasm, as were the courses put on by the union. With some exceptions, the majority of interviewees saw the union as a central feature of work and learning life at the chemical plant. One worker who was deeply involved in union activities described the learning as:

> ... it's been a good experience for me. I've been involved in a couple of things ... it was just very eye-opening how the company will take advantage if you give them the opportunity. There are companies who are a lot worse as you know, [members] are getting quite lax, you know we don't need a union in here and that, but you know if you give [the company] the opportunity. (Tom)

Of those most involved in following or taking part in union activities, there was agreement that training done by the union or involving the union was the basis for an active well-functioning union. Again, Tom puts it best, although his views were widely shared: "No training. No foundation."

Some workers commented on how union education has become more intensive (steward and executive position training in particular) over the past few years, and that this was a an extremely positive development. "I was a steward for the department for a year, but there were no courses. That was a while ago and there was a different [executive] then. They didn't really focus on education at that time. They did a few grievances ... There wasn't too much room for movement there" (Pete). Even more recently the local has seen a shift towards more emphasis on membership learning. It was said that this led to significant changes in the perceived value of the union in the workplace and encouraged higher levels of participation among rank and file members. Even those workers who did not take a particularly active role in union activities still talked about the use of courses and the use of a union in general. "[Unions] have their purpose. Like a lot of these companies would literally shit on you if you didn't have the union. I've seen that" (John).

As quoted earlier, when there was worker involvement or when workers were involved in the planning and presentation of company courses, the learning was regarded much higher in quality. Similarly, workers involved in the union and union training made comments like this woman:

> I really enjoyed the [union] training and I think they should have updates to keep you on top of what you should be knowing. When I was there [at the course] they give you all kinds of papers and ... I [found] all these different courses you can take. I'm fascinated and I want to take all these courses ... The training [union courses] was very good formal training. (Valerie)

She and other workers described the need to purchase computers for the union to gain more time for computer skills training and greater access to information from other unions. This also suggested to both researchers and union members the need for independent telephone lines for broad use by the membership in the workplace. This, workers said, would allow the local union to go "on-line" with other groups/unions while also being assured privacy from management observation. The company, partially responding to demands by workers for the opportunity to independently develop computer skills, had in fact opened up a "computer lab" for workers to use after shifts, and on weekends, and for use by family members. However, it became clear that this resource was modest (only two computers) and, more importantly, poorly advertised so most workers had never heard of it. Nevertheless, all workers felt it was the right direction for future training development.

One final issue that was interwoven with workers' learning and their union involved reflection on a difficult strike in which the local had been involved several years prior to our research. Workers who had participated often spoke of it as an important learning experience that continued to orient work and learning activities in the workplace. This strike, and perhaps more importantly the storytelling or narrative of it that emerged and was stored and transmitted to new members orally, seemed to inform and inflect management/union politics on an everyday basis. In fact, this narrative was a kind of cultural tool that could mediate a range of activity. Some used it to understand current management and union tensions, others used it as means of planning more effective workplace actions and still others use it as a more general way of understanding industrial relations. As one worker reflected,

> The learning experience around the strike that the union members felt is that they got screwed. Actually if you want to use the terms, who wins the strike – management must have won it. I don't think the issues were clear, in most places I think there was a general sort of revolt against management. Just a sort of, "we'll go on strike" without really thinking it through, and that's why they lost in a sense that today people still say, 'we'll never go on strike again.' Which is unfortunate ... because let's face it, it's the only weapon you've got. (Pete)

Conversely, as one internal union report suggested, workers may have demonstrated a degree of militant organizing that might help discourage plant changes/closure by raising the costs of doing so. In any case, especially for those who experienced the strike first hand, the learning was undeniable and circulated as a form of story that had enormous practical as well as ideological currency.

The Cultural Historical Dimensions of Learning Amongst Chemical Workers

The CHAT perspective draws our attention to several specific features of work and learning at the chemical plant; some of which have been introduced already and others that will be discussed later. One of the most prominent involves the use of OJT courses. OJT institutionalizes the classic legitimate peripheral participation or situated learning model. A CHAT analysis suggests this means that a vast range of ongoing, tool-mediated and collective participation was recognized as the means by which real learning occurs. From the company's standpoint, this also fit well with the vision of accountability required for quality certifications and offered considerable cost savings as professional training consultants or organized courses were not needed. Specific initiatives and protocols such as the company's record of employee training (used in personnel reviews), or the OJT course system (each linked with quality certification processes) suggest how tools and rules mediate a range of practices shaping specific activity systems and the way that different activity systems are linked.

The batch processing labour process in conjunction with the company's commitment to notions of worker-driven production, "empowerment" and "employee development" also contributed to many of the contradictions of the production systems being resolved by participants with the relative freedom to change patterns of participation and hence learn subtle and not-so-subtle new ways of working. This relative freedom of workers also reveals more primary levels of systemic contradictions that lead directly to those same workers questioning the appropriation of their knowledge *vis-à-vis* OJT courses opting for the development of a new form of "tool" or "instrumentality" described as a "workers' knowledge bank." This last issue also helps highlight the importance of standpoint within the work and learning activity: company standpoints aligning with certain protocols, tools, rules and divisions of labour; and worker standpoints aligning with others. Likewise, stories of a strike, or a "strike narrative" can also be seen as a tool to mediate new forms of worker-centred participation and learning. In this case, the narrative is an example of a tool that emerges from a unique (working-class) standpoint.

As with other sites, the CHAT model allows the direct integration of political-economic context as legitimate variables of work-based learning to link with our introductory remarks on firm development (Lazonick, 1991). Specifically, the sector and firm offered stable cultural communities *vis-à-vis* strong market position, strong employment, strong capital investment (each linked to domestic and international markets) in which workers' participation in various activity systems could enter into progressive and effective as well as "expansive" cycles of knowledge and skill development.

Chemical Workers' Learning and the Class Bias of Knowledge

Many factors played a role in the emergence of a strong oppositional workers' learning culture in this research site. They included a strong local union, within a sector and firm that was stable, and management strategy that expressed an open commitment to workers' learning and worker-driven production. Basically, under stable cultural and material conditions, workers articulated a distinct perspective on their own learning practices. In our report to the union local based on this research we said that these factors were key levers for workers, and important for understanding and making decisions about education/learning. However, understanding this type of learning culture requires an assessment of workers' conflicting beliefs about the nature of learning, education and even intelligence, as well as a keen appreciation for the effects of the labour process and other contextual issues.

More specifically, workers often described a distinction between "street smarts" and the "knowledge of life" as opposed to the learning and knowledge forms based in formal schooling. Many interviewees made comments such as "[y]ou can have all the education in the world but no sense" (Enzo). Others explained, "There are plenty of 'low education' guys who can learn things extensively. [There is one guy I know], very intelligent man, uneducated, but very intelligent. Education is more of an opportunity thing ... It's not really about learning" (Sean). Linked to this was the general conception of the work on the floor as it stands now. Many felt that aspects of work did not require much formal education at all. "It doesn't seem right to pick people by education, when lots of people can do the job" (Alan). "It's not rocket science but they are getting more into the computers though, so you need a bit of the computer skills, but then again, it's idiot proof ... I would say it will always be moving towards pretty simplified stuff" (Pete).

In a focus group conversation these workers outlined the difference between book-learning and "street smarts" in no uncertain terms, indicating the role of cultural assumptions about blue-collar workers:

Vince: It's like a certificate has some meaning, but a certificate doesn't denote intelligence. You know some really dumb people got the certificate ... and you know a lot of intelligent people don't finish school because it bored them. They were too intelligent for the structure that was there, they lost interest and dropped out. That doesn't mean they're not intelligent: they very well may be.

Dan: If you were measuring IQ or something that might be different, but you're not ... Like in our workplace with the office and lab people, it's like anywhere, they tend to look down on the blue-collar worker ...

Vince: You also get these people with their university educations saying, "Well what the hell could you possibly know that I don't know?" So meanwhile you stand back and say, well go ahead then, and they go ahead and build their little project and they find out it doesn't work and then they come back to ask you well what should I have done?

Dan: Actually, I find it really funny because when we're dealing with office staff or lab staff and when they're dealing with workers on the shop floor, you know, we have guys like Jim has got an engineering degree, guys like Rick who has a teaching degree, I have my real estate certificate and lots of things like that, but the people you deal with don't realize this. I'm not saying it is good or bad or whatever, but I just find it funny because in general they will look down at you because you are in manufacturing.

In another focus group, workers suggest that a preference for everyday learning is rooted in a type of cultural preference for certain types of "honest" work in which they were not required to "manipulate" people as seen among the "management scum-bags" (see example below). This cultural preference tended to denigrate the value of formal education and university degrees, but also woven into these matters is the suggestion that material resources also play a role. Workers, of course, have differing perspectives, and we can see these differences at play in the following dialogue: Ron, below, begins with a statement about why he prefers informal, everyday learning:

Ron: [It's] the only way we can afford to do our learning.
Interviewer: Do you think that's the only factor?
Ron: Yes, time and cost.
Vince: No I don't think so.
Ron: I do.
Vince: I don't think that's the only thing, it's like do you want to be a management scum-bag. If you want to be you have to at least have your university education.
Dan: Or its do you want to play along with the politics of the office. I worked as a manager at a different place once and I did not like it at all. *[pause; becomes very serious]* Actually after I ended up having to fire a close friend I decided to leave the job. I figured I don't want to do this anymore, this is crazy.
Vince: That's where some of this false sense of superiority comes from with management. "Oh well I'm university educated and you're not!" Well, hey buddy, I didn't want to go through that just to get the bullshit job you got!

Clearly there are contrasts between forms of knowledge, the cultural values related to specific jobs as well as, in the earlier quotes, the types of knowledge that are encouraged by the design of work: idiot-proof work was seen as a barrier to developing "street smarts" related to work. There is a belief that formal schooling is not necessarily the best way to learn/master production work, nor does it equate to intelligence. A variety of questions arose from these conflicting ideas. They seem to concern the following aspects of work and learning:

• Intelligence and school learning versus functional abilities: Are these the same thing and when are they different?

- Work that is simple and boring versus work that requires considerable learning and adjustment to maintain production: Did the work at the chemical plant display aspects of both with worker-driven departments (Maintenance, high performance) encouraging the latter?

- Cultural values associated with certain types of (blue- and white-collar) jobs: Are jobs based on school learning associated with forms of work and the cultural values of solidarity?

In this sense, the recognition of the diversity of labour processes and worker involvement went hand-in-hand with alternative perspectives on learning and knowledge. The phrasing that emerged among workers was the notion of "using it and *not* losing it". In order to *not* lose one's learning, in a sense, it was necessary to become deeply involved in the practical application of knowledge. In these cases, often seen in the trades/maintenance and the high performance departments, workers were able to reverse elements of the typical knowledge/power relations between workers and management: street smarts could be positively asserted. In the high performance department, John comments,

> Everyday is learning in my department … I changed the set-up for the debugging. All the "professionals" were out there, but nobody questioned how they were going to de-bug the machinery when in production. It's really about the autonomy of the department … The engineer is just the tool we use. They're too focused in their area [compared to] when you've had years and years of experience. (John)

In these contexts workers' knowledge developed rapidly.

> It took some getting used to. If you ever need anything you just ask questions. Everybody helps everybody out in terms of all the different jobs and knowledges. I still learn a lot. There was a lot of new equipment in [the high performance department] and there is a lot of new learning to be done around this. (Pete)

Struggle over Knowledge and Credentials at the Chemical Factory

In this research site there was a marked ideological struggle over what counted as learning including the class differences between alternative forms of knowledge. This struggle arose less from the type of vibrant history of trade unionism that we examined in the previous chapter (although there was a rising level of militancy in the local union), as from the types of openings that were produced by the company's perspective on learning. The company, in resisting the expansion of their "worker-driven" policies to all departments, found themselves face-to-face with a core contradiction within efforts to develop a "learning organization". Allowing workers the freedom and discretion necessary to undertake high-level, flexible, knowledge-based manufacturing work opened the way for questioning the limits of this initiative by workers who came to value their own forms of skill and knowledge production. These contradictions emerged clearly when, during our research the company set up a new warehouse facility. The company's policy of worker-driven work and design was not, initially, extended to this area. It was seen as a standard, traditional area of work that would not benefit from high levels of worker participation. However, this worker described the process of setting up the new facility as a battleground over forms of knowledge:

> [In terms of the company looking down on our knowledge] I don't find that at all; I find the opposite. I think they're intimidated. As far as I'm concerned they know that we are knowledgeable. I'll give you an example. We're going through a thing right now. There's a move from the old warehouse to the new warehouse and the manager was saying this is how we're going to move things, this is how we're going to do things [but] basically the guys [in the department] told the manager to go screw himself: Screw you, we run the place, and we're running it our way. I mean, really, that's what they literally told them, and the management eventually agreed in a way.... why can you get away with saying things like that? Because it's true and they know it ... These guys united together and basically told the manager to f–off. It really wasn't done in a crude, arrogant sense. What they were really saying is "We know what we're doing. We know how to run it." Collectively I think they got over 200 years seniority between them and they told them that. At the end of the day the plant manager did recognize what they were saying, they do have the knowledge, they do have the experience and if it isn't broke don't fix it ... [In fact] the union's role as I see it is to highlight what knowledge we actually do have and how we attain it. (Tom)

Challenging the rhetorical dimensions of workplace policies in this more informal way can be effective, although it represents a battle that must be continually fought by workers on a case-by-case basis. This form of daily struggle may actually contribute to a lively, critical and effective shop floor and union politics: it is a means of mobilizing workers, increasing participation and turning union issues into a continuous learning process. However, the preceding interviewee also points to something even more radical in terms of our understandings of learning – it involves a way of looking at learning that is collective rather than individualized. Workers in the warehouse department do not only have individual skill and knowledge, they have collective experience (over 200 years of seniority). This, as an offhand remark, actually departs from many basic individualizing tendencies within discourses of learning and probably played a significant role in the warehouse workers' ability to express their own capacities for managing their own work design.

The company thus found that there were substantial gains to be made in recognizing workers' knowledge and skill, but it did not happen without a fight. Other workers we spoke to discussed testing the boundaries of the company's policies of empowerment and recognition of the power of workers' own learning. Several interviewees spoke about the development of an expansive (non-trades) "apprenticeship" program for all workers in the plant to learn each others jobs. Workers described the resistance of the company to this idea:

> I think an important point to bring up is, at our company, we don't have any apprenticeship programs right now for the simple reason that the company, I think, they don't want to educate the people outside their area because they might just go looking for a job somewhere else. Don't move them around, eh. (Dan)

> If they officially recognize [the apprenticeship learning], they're going to have to pay for it right? They're getting the knowledge on the cheap right now. We're providing it basically free of charge. You know why would they want to formally recognize it and then have to compensate for it. (Vince)

The union was well aware of the dangers of proposing a form of "multi-tasking" (e.g., increasing internal competition between workers, making individual workers more interchangeable and expendable), they also felt confident that, if entered into a collective bargaining procedure, positive gains could be made for the membership in the form of additional compensation, the chance for workers to become invigorated by diverse challenges and empowerment of workers through a broader knowledge of the workplace as a whole.

Other incidents suggest that the company may indeed be quite resistant to having workers too knowledgeable about the workplace. A union executive recounts the following situation:

> What happened was that, there was this woman who had worked as a summer student a number of years and she was completely familiar with all the procedures and layout of the plant and everything, and when she went to apply for full-time, she was told that she was over-qualified. And the union objected and I went to see management about it just to try to get a reason why she was considered over-qualified. And it turns out that

she was studying some law course, with all due respect to people who want to learn this stuff, but it was to become a security officer ... And she was working at the race track at the time making eight bucks an hour, so I was trying to get them to explain to me how she was over-qualified if she was doing a job that paid her eight bucks an hour ... So she was caught between being over-qualified for this job and under-qualified for another. The company still hasn't explained to me the use of this term over-qualified ... At our place, which is just your standard manufacturing facility, I mean it's not rocket science, I think a lot of the qualifications are manufactured.... I mean the grade 12, is that a real qualification that you need to do the job – or is it just a barrier to some people? (Tom)

Fundamental questions such as these led many workers whom we interviewed, to question the limits of "worker-driven" manufacturing. Many questioned whether this was simply a "cultural program" to divest workers of their identity as workers, or something more. Many, in fact, went on to say that "team building," in principal, was a marvellous idea but one that should be kept quite separate from management-led quality improvement programs.

Ron: What I find, is that with this company is that they promote worker driven products, until it becomes worker driven *[general laughter]*. Then it becomes, "Oh they really know what they're doing out there" and "we can't it control anymore" ... When we have a worker driven program and they keep promoting them. It scares them....
Dan: ... But it's basically just like Ron said, you bring a supervisor out on the floor and basically they don't know, they don't run the show. We run the show. It's what we want to make happen ... Union management thing aside, you know it comes down to we control what happens out there and that's ...Our company is working towards what they call a self-directed workforce, but I think you can see that that term in itself is a joke ... Your know they say, "There's going to come the time when there is going to be no more middle managers. You're going to be responsible for your own." Which is a joke right, because you're always going to have the dragon breathing down on your head right? Let's face the fact, that if it comes from management then it's not worker-driven, if an idea truly comes from the floor – it is.
Ron: That's why when they changed the name of the foremen to facilitators - I still called them foremen because a facilitator facilitates, and a foreman dictates. You can change the wording all you want... They started calling us associates and then the reality set in and people started saying wait a minute, how can we be associates - don't we have to work together, don't we have to work as a team. So if we're not really a team, then you're the employer and I'm the employee.

Tom: What they're dealing with is psychology, eh. They're trying to take away your identity. They rob you of your identity by using the term associate and you're no longer a worker.

Dan: You more or less begin to become part of them.

Challenging Existing Structures / Introducing New Ones

Beyond the types of spontaneous, informal challenges to the rhetorical dimensions of company policy, in the course of our research with the chemical workers several organized responses to current structures were initiated. In general terms, in the local union talk quickly turned to extending genuine worker-driven arrangements beyond the high-performance department, despite much well-intentioned cynicism among many who already worked in this department. The union executive also began talking about extending the notion of "worker-driven" to the job training offered by the company, although by the end of our research the workers in this area had proposed nothing specifically. The first formal move the union made, however, involved challenging the OJT system of training rates. The union wanted more resources directed to OJT, previously used by management as a type of cost-savings mechanism. Tom, the union executive member, summed up the motivation:

> In our collective agreement we've got a training rate of 25 cents an hour which until I was looking at the research, I didn't really think about it but basically you're selling your knowledge for nothing. We're giving away our knowledge for nothing because if you look at it – if the worker doesn't train the worker then who is going to train them? The knowledge needed to do any type of job ... if you tried to bring in consultants or outside trainers to teach you ... it would cost thousands! And that knowledge we're passing on is basically free knowledge. And the company says, we should be grateful for the opportunity, and at times we've even had to grieve for the 25 cents an hour! (Tom)

This critique formed the basis for a collective bargaining proposal that suggested that the company should "put their money where their mouth was," and if they were indeed committed to OJT as a concept they should look at supporting it properly through a raise in the training rate, train-the-trainer support and a time-release from production quotas while a co-worker is training. The union was successful in negotiating these elements (e.g., a raise in the training rate) although it did not achieve all these gains and planned to return to the issue in future bargaining.

A more interesting, union-based idea that arose in the course of our research involved something interviewees came to refer to as a "Workers' Knowledge Bank". It emerged in the course of a focus group discussion that dealt with Prior Learning Assessment and Recognition (PLAR); a system of assessing and recognizing the informal learning that people do outside of classrooms. What was most interesting is that these workers recognized real limits to how such PLAR systems could be used in the workplace (see Sawchuk, 1998) as management ultimately controlled what knowledge could be applied and developed.

> From my point of view, from the workers' point of view where I'm coming from, I don't think it's too practical. I mean it's good to write everything down, but as long as there are two sides drawn between management and the working class a lot of it is not applicable. They just are not going to listen right. I'm not trying to be hard-nosed about it, and a lot of it I think is great for among workers, but as far as management and workers coming together and people saying, yeah we're going to listen to you. (Dan)

Discussing the vast range of general learning interests, work-based and non-work-based, that was described briefly in the profile section earlier in the chapter, these workers began to think of ways to store, share and develop their knowledge as a union local.

> *Ron*: What I find is that among the bunch of us here we'd be willing to help one another outside of work, or on the job – we'd be more than willing to help one another. I have skills that maybe Bill doesn't have. Frank has knowledge that maybe I don't have – we draw on each other I guess … but when it comes to the workplace and management and if you are being resisted for the knowledge you have – you're not going to give any to them! … *[Instead]* I know that my neighbour or a friend or a co-worker has the knowledge and I can draw on it because, they're going to help me more willingly than if I was wearing the badge "Manager" or wearing the *[supervisor]* Blue-Coat.
> *Tom*: Like a knowledge bank! Like a knowledge bank for workers internally. *[speaking to the group]* You know it's like Lou is a dap hand at mechanics. If anybody wants help winning a grievance – give me a shout! *[general laughter]*
> *Dan*: Or Lionel with the camera there!
> *Tom*: Or even collecting it in a whole workplace or even workers in general. I can see a lot of value in that. I didn't really think about it until we all started talking about it actually *[laughs]* …

Tom returns to the notion of the Knowledge Bank later in the session and explicitly combines it with the idea of PLAR as tool for extending and

fortifying existing cultural networks and impulses of workers to band together to share (class) experience and develop understanding.

> To me the best learning you can have is the informal learning, as Ron
> called it the education of life ... [An] example is as simple as a guy came
> to me last week about his friend who doesn't work at our workplace and
> isn't unionized, and he asked me a simple question about the hours of
> work the guy should be working ... so the education that I've attained
> through the years – which isn't hard to learn – I basically tried to pass on
> to him which opened his eyes ... So what started as a simple informal
> question, I tried to develop it into a sort of broad education for him to
> think about at least. Maybe even do something about. But the point, is
> you learn everyday, and you teach everyday ... In the labour movement
> this like a tool and it's an excellent one. We've sent people to many
> courses to acquire some formal and informal skills or whatever, but this
> [PLAR] would be a good tool to keep a record of their learning, person-
> ally, and for a local ... And then you can get into things like electronics,
> and car mechanics, painting and decorating, etc. You could sort of build it
> out from there. But all with the idea that you pass the learning on. (Tom)

The research process of exploring, together with workers, the power of
their existing everyday learning networks, in conjunction with ideas like
PLAR and the Workers' Knowledge Bank, helped us all to see practical
ways that labour education can be supported and strengthened at the level
of the union local.

Concluding Remarks

This chapter has emphasized the themes of an oppositional and vibrant
union learning culture amongst a small, but relatively secure and well-educated membership of (mostly) middle-aged, white men. Basic sensitivity
to understanding the socio-cultural, historical and political economic dimensions of learning in all their forms has allowed us to more clearly dis-
cern the linkages and the struggles over work and learning that workers
like these face.

Specifically, the workforce is composed of people with a range of industrial experiences, confident of their abilities and vigorously interested
in guarding their rights and expanding the potential for more satisfying
work, greater control and an opportunity for greater contributions to production. In fact, we saw how an active union local can challenge the rhetoric of management on the issues of "worker-driven", high-performance
production in a "learning organization". Indeed, in some cases the local
has used this rhetoric effectively to make gains for their membership in
terms of autonomy, work and learning. Moreover, a range of creative re-

sponses, such as those related to the Workers' Knowledge Bank, have emerged and suggest valuable options for this and other groups of workers for future proposals. The political economic context and firm characteristics are inseparable from the activity and its future trajectory.

Notes

1. The term "associate" is used by the company and was introduced into the collective agreement just prior to our research, however, by the end of our research the term had been removed and the term "employee" was replaced based on the union's urging. This was felt to be an important point of maintaining worker identity by the negotiating team, and although not a "strike issue" was not removed from the bargaining table despite the company's insistence.

2. QS certification parallels the ISO certification, but is unique to those supplying products to the automotive sector.

3. The ISO and QS certification programmes state:

    The supplier shall establish and maintain documented procedures for identifying training needs and provide for the training of all personnel performing activities affecting quality. Personnel performing specific assigned tasks shall be qualified on the basis of appropriate education, training and or experience as required. Appropriate records of training shall be maintained. ("ISO 9001: 1994 (E)," ISO Standards Compendium, ISO 9000 Quality Management, 5th Edition, Geneva 1994, p.153 as cited in John Anderson and Jonathan Eaton. "ISO 9000: An Examination of the Labour Issues in the Implementation of ISO 9000 Certification" Labour Caucus for Electrical and Electronic Industries, April 1995) and (International Organization for Standardization (QS-9000, 1996; 3rd printing).

4. Beyond the collective agreement, this training was also supported by the company's quality steering committee's "Quality Plan". This committee has committed in principle to 40 hours of training per employee per year (1995-98), but again with the use of "OJT" training the 40 hours commitment does not indicate just how actual resources are allocated.

5. Training is in fact the seventh most common reason for QS-9000 non-compliance,

    Assessors often use the training files of new employees as a starting point for evaluating a company's compliance with training requirements. Were the new hires trained at all? Were they trained within the company's set time lines? In addition, companies may neglect to evaluate and document the effectiveness of training. It is also important that your employee's qualification match those listed on the job descriptions. If they do not match and the discrepancy is not explained, it could result in a non-compliance. (Bakker, R.M. "Top 10 Barriers to QS-9000." Actionline, May 1996)

6. The maintenance department by its very nature, i.e. they come in when there is a problem – especially in light of updated technology from time to time – requires a good deal of active problem-solving and judgement. However, many workers (in all departments) often described "fixing" their machines/equipment themselves.

7. Different workers in the high performance department described the work in different ways. These workers detailed chances to exercise important judgement and problem-solving, but that the basic work, at times, could still be straightforward.

CHAPTER 5

Learning, Restructuring and Job Segregation at a Community College

Jackie Harris is an office clerk at the college. Growing up on a family farm in Southern Ontario Jackie learned that the "education" a person had said little about their capacities for learning. Politics and social justice were discussed nightly around the dinner table.

A critical perspective on life, learning, and the basic capacities (indeed responsibility) of all people to contribute became a life theme, but as a young girl Jackie learned other lessons as well. At school, she learned that "girls had their place." Nervous to break out of this unwritten code, Jackie remembers how her hand would shake as it stood outstretched to answer a question, even when she knew the answers. She also realized that the most important learning came out of community:

> *[In life] we need fewer teachers and more students. You learn a lot from a group without having to have an expert tell you everything ... but there has to be a community where ... you are accepted, where you are not judged, where they can ask stupid questions and where you can learn about what's important to you ...*

And, growing up Jackie also encountered the world of problems of work, talking informally and excitedly with co-workers about organizing a union at the local fast food restaurant where she had a job for a time. Her family could not easily afford to send her for post-secondary schooling but a small scholarship eased the way.

With her husband working on and off, she began a series of jobs, first as a day-care worker in a low-income community in the city, shortly af-

ter obtaining a job as a visual artist for educational initiatives at a downtown venereal disease clinic. Here too she confirmed the importance of people banding together in communities to battle difficult circumstances. Jackie eventually landed a position as a clerk in the mechanical department of the college, learning from co-workers as she went. She eventually was able to do most AV repair work needed and was hired full-time despite, again, feeling as if breaking an un-written code that this was not "women's work." Throughout this period and into the present, as Jackie says, it was and is tough living as part of the working poor. "My friends, all underemployed, from my generation. [Lots of us] have university degrees but are maids or waitresses. Everybody I know. If not degrees, they have incredible intelligence ..." These circumstances and her background flow through Jackie's learning and work as a union activist at her workplace where she now works as a clerk (having been bumped to other jobs around the college). The union work and the extensive learning she does to keep up with it keeps her busy while it also imposes tensions in her home because she's away so often. Leading several union committees, in the last year she's also completed her bachelor's degree, and here again it was the support of co-workers that saw her through. Jackie informally "partnered up" with another woman doing the same thing. She talks of how they shared babysitting and, in general, helped each other with their studying late into the evenings.

When Jackie talks about her workplace she cites job and training "ghettos" that short-change, isolate and narrow the prospects for many workers. Based on her learning life, she advocates more time for people who do similar jobs to get-together: "we need buddy systems." She references massive cuts to resources everywhere on campus (except for "shagland" – the name workers give to the executive management offices because of their lush shag carpeting), and she cites instances of open racism, "tokenism," sexual harassment, and the arbitrary rules of training access as degrading the learning potential for the majority of employees at the college, she adds, working in the public sector is inherently contradictory.

Jackie's life is an extreme example among workers at the college in several ways but not all. Most of the workers we spoke to at the college were engaged in extensive learning efforts largely on their own initiative; virtually all interviewees experienced the contradictions of restructuring and work intensification in ways similar to Jackie with many citing the same types of barriers either directly or indirectly. Jackie Harris's story also points toward some solutions for her workplace. She

*and many other workers learned early on in their lives that people work-
ing and learning together as valued members of a type of community and
integrated as legitimate decision-makers in that community are often the
best way to respond to times of change and difficulty.*

Introduction

The community college that we chose for the study is one of the largest in
Canada; with many different campuses that provide services to a diverse
cultural community and a combined enrolment of over 50,000 full-time
and part-time students. The college offers a core set of academic programs: Applied Arts, Business, Science and Technology. It also offers
trades training funded by the provincial and federal governments, and ad-
ministrates continuing education and distance education departments. The
vision that leaders of the organization orient towards is that the college is
a centre for opportunity for people from every realm of society (College's
Annual Report, 1994).

At the time of our research, the college had been reorganizing the
workplace to streamline its services in the face of severe budget cuts and
public sector restructuring initiated by its main funder, the provincial government. Funding levels dropped throughout the early 1990s but under the
direction of a right-wing government beginning in 1995, things worsened
as austerity measures, combined with the government's belief that colleges should generate more of their own revenue, created the need for
large-scale reorganization in the latter half of the decade. During the
1995–97 period of our interviews, the total transfer grants from the provincial government to the college system dropped by about 30 percent.
The pattern of funding distribution for this particular college was similar,
if more severe. Like other colleges, attempts to make up some of the short-
fall included raising student tuition (12 percent of total revenue in 1992;
30 percent in 1997) and by seeking out other means of generating revenue
(28 percent of total revenue in 1992 and 45 percent in 1997) such as private training contracts. Total revenue still (in 1997 dollars) dropped by
$13M between 1995 and 1997 while revenue generation (beyond simply
raising tuition fees) also incurred extra costs. Given the unlikely prospect
of changing the government's mind, some form of serious restructuring
had to occur in the college.

On the surface the public sector in Canada seems less bound by direct
market forces and more inclined than the private sector to make employment policy decisions on the basis of fairness, equity and the "community
good" (Lowe, 2000). In the 1990s, however, the public and private sectors
shared the strategy of downsizing and wage freezes. In fact, public sector

employment levels shrunk more than any other industry's with the 1996 census marking the first drop in the absolute number of (federal and provincial) employees recorded since 1901 (Lowe, 2000: 130). Although the college is not a direct governmental service provider, at the time of this study there were around 480 full-time support staff and 720 full-time faculty members at the college, and this number continued to drop as the study proceeded.

Our study did not survey the work and learning practices of faculty members or management personnel, but focused on the most vulnerable staff, namely the many different types of clerical and other office support, technical, security, shipping/receiving, maintenance and custodial staff. The two largest occupational groups were the clerical and office support followed by the custodial staff. Of these workers, clerical staff primarily comprised women workers within the age range of thirty to forty years, with relatively few older workers. These people contributed to program coordination, registration, secretarial and other administrative functions. An even mix of men and women made up the custodial staff, while the technical and maintenance staff categories were largely filled by men. Diverse cultural backgrounds are represented throughout the workforce with people predominantly from Mediterranean Europe, the British Isles and the Caribbean working in support staff positions. The vast majority of the workers who are first-generation Canadians work in the custodial and cleaning department while second-, third- and later generation Canadians often occupied the clerical, technician and other college staff positions.

The selection of interviewees was based on gender, age, seniority, formal educational background and occupational group (see appendices). The interviews were carried out between September 1995 and January 1997. Research at the college was based on interviews with twenty-four workers, fourteen women and ten men. The participants ranged from twenty-eight to sixty-three years of age (average age of forty-two years) with seniority ranging from one to twenty-seven years (average seniority of eleven years). The staff members represented in this research have a wide range of formal educational background, from those had not completed high school to those with advanced university degrees; over 60 percent of those interviewed had some form of post-secondary education, although differences depended greatly on occupational group. Many of the clerical and office staff interviewed were currently involved in some type of course-based skills upgrading, while others had completed formal educational certificates, diplomas, or degrees while working at the college. The report integrates interview material from all of the participants. Job titles are indicated where relevant when direct quotes are used; however, for the purposes of confidentiality personal information has been modified to assure anonymity.

In the introduction we described the college site as involving a mixed "adaptive/innovative" managerial strategy (Lazonick, 1991) in which degenerative approaches such as work intensification are mixed with an innovative rhetoric, and to some degree innovative practice, regarding employee development and "participative management." The result is mixed experiences of learning opportunities based on occupational segmentation. The college seems to be a rich, if segmented, environment for learning opportunity. The labour process involves increasing rationalization of outputs and accountability, along with intensification and multi-tasking. At the same time, the organization's decentralized structure contributes to a labour process in which people, despite the challenge of dispersed campuses, still seem able to see how their work fits into the overall organizational structure.

Working at the College
Increased Workloads

Restructuring at the college has resulted in increased workloads for many staff, largely due to merged jobs, decreased staff members, and increased numbers of students. It is clear in talking to workers that this intensification has become difficult to manage at the individual level. Men and women in the custodial area are also feeling the effects of increased workloads as well.

> We have three shifts and six phases in the building. There are four floors. Five years ago there were fifty full-time and seventy-two part-time. Now there are about thirty-five full-time and four part-time. So the areas are larger and they expect the same quality of cleaning. That's impossible. And at the same time there are four to five thousand more students. (Ken)

> More work and less guys spell trouble in this area. (Scott).

One answer to the problem would simply be to hire more people to absorb the extra work: "One thing they could do is not to ask us to work overtime but hire others to do the work. I guess sometimes they don't have time to train. Yes, if there was a pool of people whose training was kept up, that would work" (Ursula). However, it appears that not all departments are facing the same shortage of workers. For example, the computing department, which is a priority for the college, has been supplied with additional help, while other departments, like admissions, have not.

Loss of Control Over Work

There is a feeling among many staff members that it is difficult to think about training and education at this time when people are not sure what kinds of jobs they will be doing in the near future. Several of the interviewees felt they were either on the verge of being let go, or they could be bumped into an entirely different job. Very few people, even those who felt secure in their present jobs, were confident that they would be able to find new jobs. This situation has had significant negative effects on people's lives, and on their capacity to plan for change:

> People are feeling uncomfortable and insecure ... [workers] are not interested in doing things, they're interested in sustaining things, through systems and stuff. But they're dying. Everything around them is dying. (Bob)

> You're not in control of where you want to be. You spend money, go to school, get an education, work for years and you're still not in control of where you want to be. You apply all over the place and everyone is in the same situation, the available jobs are flooded. Or you are bumped to positions that have hiring freezes. I don't know ... (Sara)

Perhaps because of their anxiety over potential job loss, many of the suggestions for learning centred around job training. Several staff members have experienced job transfers in the restructuring that began in the mid-1990s. More than one person had been moved from a job appropriate to their prior job training, to a totally unrelated job. In other cases, people have kept their jobs, but have had additional work "downloaded" onto them. For example, one person (Florence) is now doing her current office clerk job as well as her old receptionist job in her area. Others are doing what previously had been part of the faculty job description (the non-teaching functions), and in some cases doing significant amounts of unpaid overtime work, sometimes two to three hours of (unpaid) overtime per day during peak times.

Restructuring has made it difficult or impossible for some people to return to jobs they are interested in or are qualified to have. Some displaced workers have been faced with lower waged/lower skilled work. In other cases, testing procedures for placing people in new jobs have unfairly screened people out of jobs within their pay-bands. It is possible that the impact of restructuring has been more severe in some job categories, such as the caretaking and registration/clerical areas, because more people can qualify for such posts, while fewer workers are able to apply for positions in technician and maintenance areas due to credential requirements.

Learning at the College
Job-Based Training at the College

The formal training (of the workers interviewed) includes certificates or diplomas in the service industries, recreation, technical, commercial and secretarial training, Bachelor of Arts and Master of Education degrees. While interviewees at the college site held a variety of educational credentials, most of the people we talked with did not have special training for some of the work they are now doing and rely mainly on co-workers, professional development courses, or take courses outside the college in order to stay updated on things they need to know for their jobs. Most people have taken advantage of some kind of formal work training on their own initiative. Although professional development is encouraged, there is often little time for training in the current context: "The colleges' attitude is ... here's your system. Do your best to learn it. Here's some PD [professional development]. Take it if you need it, and ... boom! You're on your own" (Lacy).

In general terms, the college setting would seem to offer extraordinary opportunities for workplace learning among workers and most interviewees confirmed this. There are a variety of programs and opportunities, but what is more unique to the college setting is that learning is both necessary for work and a "product" of work in that the workers run an educational institution. These features all contribute to the fact that work-based learning, perhaps more than in our other research sites, seems especially valued among most of those interviewed. A basic profile of the formal schooling and further education background for the different unionized workers in the college setting appears in Table 5.1.

Table 5.1: Participation in Formal Schooling and Further Education / Training

Occupation	Formal Schooling	Further Education/Training
Administration Officers	High	High
Computer Technicians	Medium/High	High
Maintenance	Medium	Low
Clerical	Medium	Medium
Security	Medium	Low
Shipping/Receiving	Medium	Low
Custodial Lead Hand	Low	Low
Housekeeping	Low	Low

Those with higher pay, status, authority and organizational responsibility tend to rank higher in both formal and nonformal learning categories. This distribution reproduces the historical, "them who had are them who gets" occupational, class-based pattern of participation in education. There is a general trend of credential inflation in the majority of occupational categories. Many workers outlined how they have formal educational credentials necessary to obtain the job but largely superfluous to actually doing it. Custodial Lead Hands as well as Housekeeping (hand cleaning) workers, on the other hand, tend to be under-credentialized, at least in terms of educational credentials recognized here in Canada.

In terms of formal learning undertaken during the period of our study, we found that staff can sign up for a limited number of courses (one per year) from their own college. Course and registration fees are reduced by approximately 90 percent. However, only six of the twenty-seven workers we spoke to reported having taken advantage of these courses within the year previous to their interviews. These people indicated that they were able to do so because the courses were affordable and easily accessible in terms of timing (i.e., many courses are available at night, lunchtime, or weekends) and in terms of the course locations (i.e., at their workplace).

> There are advantages at the college ... A couple of guys [in the custodial area] are taking courses. One guy has only a couple more courses to get his building operator. It took him two and a half years. Still, ten dollars a course is nothing. It might take you twice as long, but you're not paying full tuition. In the meantime you're working and getting a pay cheque, and getting a degree or diploma. (Ken)

In addition, the employer had arranged access to continuing education at selected provincial universities at both the undergraduate and graduate levels for their staff. However, these courses are not offered at reduced rates nor were they supported by time off work: "I have taken courses in order to be on the cutting edge of things ... I have done a lot of learning, and spent a lot of time and money done on my own" (Alphonse).

Not all the course-based learning contributes to a formal educational credential and, compared to other sites in this research, the college setting is arguably the most developed in terms of the employer's course-based training offerings. Central to these offerings is the college's Centre for Employee Development (CED). The CED provides in-house workshops and other services including career counselling and aptitude testing. Its mandate is to "facilitate lifelong learning and provide training and professional development to the college's faculty, support and administrative staff."

Most of the workshops are available at least once during a college term. Workshops offered during our research were computer software learning, personal health and, more recently, job preparation skills. There were over forty courses offered in winter 1997 with the majority involving computer applications. The CED currently organizes their courses into Organizational Development; Personal Development; Teaching and Learning Effectiveness; Continuing Employability (Career Development); Technology and Computer Applications in Administration, Teaching and Support; and Accountability. At the time of our interviews these workshops had, in fact, been shortened significantly (from an average two day duration to a half-day) provoking much criticism, e.g., "they're practically useless," from workers.

While almost everyone we spoke with mentioned that computers are something they think they should be learning more about, this focus is not necessarily what people would prefer to be doing. According to one staff member, "My heart isn't in computers because it's something I love and want to learn all about. It's because I know there is a job there in the future." At the time of the study, learning to use e-mail was widespread though mediated by access to computers distributed by management. Computer learning, in this sense, was a key form of course-based learning that involved both voluntary and compulsory participation. In other words, it was sometimes an interest of workers but it was almost always a skill necessary for keeping a job or getting a new one.

The CED also makes general interest workshops available to staff. However, it emerged that very few workers actually took these broader courses despite the widespread general interest in them. The uneven distribution of access to course-based learning among the unionized workers is discussed below, but basically, although the Centre is open to all college employees, there is far more training made available to (non-unionized) faculty and management personnel. As a unionized worker explained, "they have more varied needs" (Alice) that accompany seemingly more complex and more important work that is apparently more valued by the college. Our interviewees were most likely to have taken workshops in the areas of Computer Applications, Personal Development and Continuing Employability. The courses that people take generally depend on the type of work they are doing, although most people reported taking computer software and effective communications workshops.

The college views professional development as a shared responsibility between the college and its employees, where the college's role is to invest resources into services and to provide learning opportunities for employees, and where employees are to take advantage of workshops, find ways to apply their skills and knowledge in the work process and play a role in suggesting new ways for the Centre to contribute to training and productivity. However, it is crucial to recognize that in practice workers are responsible for negotiating access to training with their supervisors on an individual basis. This negotiation revolves around unilateral managerial decisions on the relevancy of the training to the job, the contribution of the training to career development of the worker and the manager's decisions over the distribution of departmental training resources among specific workers – time off must be granted and workload must be re-distributed. While on the surface professional development is voluntary, in some cases it can be mandatory as the result of supervisory performance reviews. Most workers who were required to engage in professional development had to take workshops such as Effective Teamwork in a Participative Environment. Thus, the training system is one with considerable biases, all of which are weighted towards managerial imperatives, departmental performance and managerial decision-making, rather than towards workers' sense of their own interests and needs. The rhetoric and resources suggest a rich, open training environment, and there are certainly extensive opportunities for worker training, but learning is also shaped by the distribution of power among managers and workers. Thus, training at the college represents a contested terrain of learning and organizational control.

One dimension of the work-based training process that may be fairly unique to the college site is the opportunity for workers to both learn *and* teach. In some cases, unionized workers can become part-time instructors based on the general levels of expertise (no matter how attained) with these tasks becoming part of their official job description. These instructors must be supervised and cannot, officially, evaluate students, but amongst these workers it is felt that this provides support to expand their knowledge which in turn contributes to their own work-based learning. Opportunity for participation in this way emerges largely from informal contacts with college faculty (i.e., jobs are not posted for unionized workers), and in this way is governed by the general networks of communication and participation within the college as a whole. Workers with specific mechanical or computer-based skills are most likely to engage in these teaching/learning experiences. One worker explains how he became an instructor and was able to use professional development courses to enhance his teaching:

One of the instructors said to me that I had a very good way of explaining things and that I should teach so I wound up teaching a couple of subjects for several semesters ... As part of the on-going professional development [this encouraged me to take] an Adult Training certificate, which gives you techniques and methods and understanding about giving instructions. And I took that to help make my teaching and lessons more effective and to give myself recognition for my abilities. (Frank)

However, there are some problems with the worker-as-instructor role. Not only are the majority of job groupings left out of the process, but recently the college has moved to download more teaching responsibilities into lower pay categories. There is a concern that this move to create "para" professional jobs may lead to tension between faculty and unionized workers around the distribution of work, and may further widen the gap between those who design and those who deliver the curriculum.

Informal Learning for the Job

As in most workplaces, workers report that they mainly learned "on the job" with the help of co-workers. Not surprisingly, people tended to report different specific means of accessing information and learning, depending on their job. For example, the computer and media technicians reported doing much of their most important learning by reading manuals and trying out the equipment. Custodians typically learned their jobs in a traditional style of apprenticeship based on watching and working with more experienced workers. Informal and self-directed learning was equally important to the other occupational groups:

At first I worked with someone else and was shown the job. I learned from him, but about 70 percent by myself. It was very stressful at first and I am still learning something every day ... (Lisa)

When I first took the job, it was a new position. There was no one to help me because no one had a clue. I was given no direction so I sort of made the job up and learned as I went along. I figured out as I went along who I need to get in touch with; how I could create some space for this activity. (Ellen)

Computer knowledge has become important for most jobs at the college and most workers have spent considerable time learning on their own, either at work or at home (using self-guided, computer-based tutorials). One staff member explains how he picked up computer knowledge on the job:

I was completely computer illiterate and I've become a little better, I've become a lot better just by doing it, by necessity, I guess. I don't think

I've ever read or gone through a tutorial on any of the software, but I tend to use what's applicable to me. So, if I need to do a memo, I learn how to use [the computer], and do it. Once I'm more comfortable with what I need to do, I tend to expand from there. (Marco)

Self-guided reading was noted as important to certain jobs, especially the media and computer technician and coordinator groups. For example, one woman subscribes to a computer magazine in order to keep up with the information she needs for her job. Some staff rely heavily upon the college library and take advantage of periodicals circulated by the library to staff members.

Amongst all workers, the interview data show clearly that staff members' work experience is as necessary as, if not more necessary than, any diploma or trades certificate of qualification when it comes to actually doing the job, so much so that most interviewees indicate resistance to the separate discussion of different forms of learning: "You take information from every source. You can't segregate it" (Frank). The ability to work together and communicate with other workers was viewed as the key to learning in areas like technical support:

Not only do you have to have skill, you have to be able to work as a team, you have to be able to communicate. And I think the group has been together for a long enough period of time. [We're] well blended. Basically, one of the things the lead hand has tried to promote, and I think is a good idea, is the ability for each person to do some of the other person's work. (Frank)

In areas where work has been recently intensified, informal learning has involved trying to manage things like workload. Some staff members, for example in clerical work, have tried to address the problems of restructuring by talking about these issues outside of work time: "Almost every day there are discussions at the lunch hour and after work with workmates … there has been a lot of learning in terms of discussing with co-workers ways to manage our workloads" (Cathy).

Workers said that they learned informally based on a web of experiences from participation in a variety of settings within their current jobs. Interviewees reported drawing on various skills gained in areas like teaching, community events, volunteering, or being a parent, to apply in their work. For example:

You need to get along with the students in my job because they're with the faculty all the time and sometimes they let off steam to me. I just listen. You don't repeat it or anything. It is very important to get along with the students. I learned to get along with my kids. I learned it as a parent volunteer with the school where some of the kids didn't want to learn, to read, or do anything. I think that helped. And I think with my boys and their friends coming in the house all the time, I could get along with them. (Maureen)

Making use of a complex web of informal learning seems to require a broad understanding of how different work-based sites fit with each other. Maintenance workers, for example, said that learning about other jobs was important for gaining a broader understanding of the system they work within. This was true for both specific job skills and for learning about the college administrative structure. This knowledge was referred to as "overlapping skills." For example:

> I have learned to do things that are not part of my trade ... A good thing I have learned here is about fire alarm systems ... which is related to building vents. I have learned this from someone who has been there a long time, and who is a licensed electrical technician. With the building system, generally one learns as one goes along, from the other guys. (Frank)

Learning informally from workers *outside* of the college was also viewed as an important means of gaining job knowledge. This staff member explains how outside expertise might be distributed amongst workers:

> If I'm not occupied with something else ... [the vending company technician] will tell me what they are doing when they're fixing it. And we'll talk back and forth and I will learn what I can about the system ... And quite frequently later on if there's a problem and one of the other staff is working on it, then I'll explain about what I know. (Marco)

These types of learning linkages to organizations inside and outside of the college also serve an important function for a range of other college workers. The linkages were formed by, for example, telephone contacts to people outside the college (vendors, other colleges, etc.) attending conferences, and volunteering on various non-college organizational boards in the community. Through networking with a host of college-based and outside contacts, college staff have sometimes been able to learn, or imagine, new ways of organizing their work. These kinds of broader visions of work are a form of organizational knowledge or meta-workplace learning, and can contribute to making sense of the sometimes fragmented experiences of informal learning in their specific departments. As such it places immediate work-based informal learning in a broader context. For example, worker involvement in contributing to college policy as members of various workplace committees such as the administrative performance review or professional development committee can make an important contribution to work and learning by contextualizing it more broadly.

Union-Based Learning

Learning for the purposes of performing one's work did not comprise all forms of learning among the interviewees. About half of the twenty-seven workers we interviewed had been active members in the union at one time or another and they reported a variety of other significant forms of learning. People said that they became active in the union for a number of reasons, all of which depended upon their involvement with various social networks. For example, many were encouraged towards union involvement by friends, family members, even workplace suppliers and co-workers who were active in a union. According to interviewees, restructuring had led to a levelling-off of union activity with fewer people "willing to stick their necks out." This negatively affected the role of the union in member learning although it was partially balanced by the fact that the general membership has had more direct contact with the union than normal because of job inquiries. During our study, the union local had increased the amount of resources directed to its own forms of training activity. This complemented the standard set of tools and issues courses (e.g., effective stewarding, the grievance process, and health and safety) offered to activists and included courses in things like labour history taught to local members through the city's Labour Council.

Interviewees said the union courses have been important sites of learning for both personal and organizational life. Most said these course contributed to "activating" activism in the workplace and beyond. Other workers recognized their past union activities, especially their hands-on experience with organizing, as important for gaining skills that they use in their regular jobs:

> The union has given me excellent skills in public speaking, how to run meetings, minutes of order ... The training unions provide is more in-depth [than other employer training] ... The kind of things you learn are great. Like legalistics. They teach you how to lobby and how the politics works. (Lacy)

People also indicated that union-based learning sensitized them to issues relating to their work at the college, like student scheduling problems, diversity and disability issues. The union courses have also given some people the experience, interest, and self-confidence they need to pursue other formal learning activities, including courses at the college.

Union education has also taken place informally through attending meetings and sharing information with other workers. Two union representatives mentioned that union meetings and educationals enabled them to learn more about workers' experiences in other organizations around the city. Union meetings have given people more direct information about the work situation at the college generally making it an alternative form of broader, "meta-learning" mentioned above.

The general membership seems to rely mainly on union mailings to keep up with what is happening in the union; most notably the status of the collective agreement.

> I don't get very involved in it. I read the notices that we get, I read the memos that we get. I think that I've probably been to one union meeting. I'm not a strong supporter, but I'm not a dissident, either. That's the situation I'm in now, the union position. So, I keep up on what's going on. (Connie)

This type of union-based communication structure, in fact, provides the foundation for yet another important strand of the overall web of learning in which workers engage. Indeed, there appears to be some support for this type of union-based learning approach among those who were not active in the local. For example, some staff members would like to see more union involvement with a broader range of issues affecting the membership:

> People should play more of a role in the union. Most of us just know about grievances – but what else is the union doing? Guys have been fighting to get classifications changed but it should be more involvement, more friendship. I really don't know what else the union does. (Jean)

The need for education about the union was pointed out by more than one staff member. One worker who has not been active in the union is concerned that everyone should be more informed about the role of the union, especially during restructuring, and that management is quite defensive concerning issues of unionism: "I find some management people are so against the union they forget their role is to be honest with us. Their fear of the union having a power they don't like stops some of that information flow so that we don't learn the same way" (Alice).

The Cultural Historical Dimensions of Learning

The CHAT perspective draws our attention to learning as a broad, critically situated field of participation, and in the college setting specifically this means integrating into our analysis the unique forms of public sector restructuring as more than simply the context of learning. In general terms, our profile of learning has already laid the groundwork for this by providing a detailed discussion of the changing distribution of resources, the changing conditions of work and in turn this allowed us to better understand the structure of people's participation with each other. This allows an interegrated CHAT analylsis for a deeper understanding of workers' learning. In other words, what conventional learning theory views as mere "context" is understood as inherent to the learning process.

In the following sections we see how activity consciously aimed at "learning" is mediated by a particular set of institutional discourses, e.g., "professional development," particular organizational spaces, e.g., the "Centre for Employee Development," and particular developmental tools such, e.g., "Personal Professional Development Plan." These tools embody particular historical relations and offer certain material and normative affordances while inhibiting others. Hence, learning among college workers is seen as an extension of an entire, historical system of activities that constitute working life at the college. Our discussion of the changing divisions of labour, *vis-à-vis* multi-tasking and work intensification relates to the segregation of groups of workers. Finally, we highlight the specific forms of organizational contradiction that have emerged with restructuring, which, in turn, help drive change. The concept of "contradiction" is central to CHAT analysis, and at what is known as the "primary level," these contradictions emerge in the form of austerity programs that relate directly to the expanding reach of market principles and commodification into the public sector and workers" learning lives. This primary contradiction, in turn, expresses itself in a variety of ways locally: shifting of responsibility for training onto the backs of workers, the temptation of management to hyper-exploit particular groups of workers, etc. However, the contradiction also reveals itself in terms of the reactions of workers to draw on activity that is outside processes of the formal labour contract and economic exchange, as when they involve themselves in informal learning and other social networks – whether the networks involve co-workers or other people in non-work situations.

Restructuring, Work and Learning in a College Setting

The basic profiles of learning help us to see the range of practices and structures in place at the college but also put us in a position to better understand the specific forms of contradiction and emergent patterns of work and learning that accompany restructuring within the sector. Although there are important similarities, the public sector is distinct in some ways from the private sector in the way it approaches restructuring. For one thing, in Canada the public sector has a much higher union density, which affects the culture of decision-making. But perhaps more importantly, although subject to its own version of "market-discipline," government austerity does not seem to involve the exact same types of changes as does the profit imperative of the private sector. In the educational sectors specifically, these factors are mixed with, as mentioned above, a view of "learning" as an intrinsically valuable activity. This informs how management and workers deal with one another in responding to restructuring – specifically, restructuring in the college setting as it is shaped by greater unionization, a heightened faith in learning as a means of responding to change, and, in turn, a tendency towards strategic planning that makes use of progressive rhetoric, policies and structures. Nevertheless, in the reality of practice, a variety of contradictions still emerge. Scarce and inappropriately distributed resources as well as unilateral control over training access expose gaps in the managerial rhetoric and planning when viewed from the standpoint of college workers. Moreover, the terrain of learning becomes governed by the processes of "bumping" (workers being moved around), downsizing, multi-tasking and work intensification.

The Participative Management Strategy

Alongside the specific programs and policies of the CED, at the heart of management's attempt to restructure in response to decreased funding was the "Participative Management Program." This program was adopted at the college to encourage worker input. At the beginning of our research, the college's Annual Report introduced the Participative Management program: "During 1994, the process of creating a more empowered workforce involved nearly 70 percent of college employees in Participative Management workshops with their work groups. These sessions are designed to incorporate real issues that work groups currently face, allowing them to solve problems or challenges with which they have previously struggled." However, one barrier to worker skill and knowledge development noted by several people has been the failure of this program. Although Participative Management theoretically offers workers the potential to make more decisions about their work conditions, there have been limitations as to what role people are actually playing: e.g., "They're giving input and ideas and the actual decisions still rest with the board of governors" (Jackie). Implementation of the philosophy seems to rely on individual managers and their previous relations with workers, fuelling the uneven distributions of training opportunities introduced in the previous section. Some department managers have followed through on the concepts while others have followed a traditional top-down management strategy. Most workers said that they have not seen much coming out of the Participative Management program:

> I don't think they educated us on Participative Management because now that the changes are happening we're not being told anything ... I don't want to sound like a bitter employee because I don't feel bitter. I've been

thinking a lot about it, that there's a difference between giving people the impression of something and actually teaching somebody and allowing them to learn what it is they're dealing with, as opposed to just trying to explain a process to them. And I think they put a lot of energy into explaining a process but we didn't learn about it. I felt that that's what was missing. (Alice)

The way things are now fosters a competitive spirit – people are afraid to depend on others for fear of being seen as weak – decentralization of power is not what is happening here. One person is in control. (Richard)

In general, staff members suggested that genuine participation in the college administrative community would have positive effects on themselves personally, on the individual departments, and on the overall college structure: "The whole idea of empowerment, if you give the average worker more responsibility, they feel more responsibility towards the goals of the whole ... If we're given more autonomy, we tend to be more creative, too" (Frank). And although there were examples where workers had made a significant impact on the management decisions in their work areas, there were more instances where ideas were not taken seriously. One person described being shut out of important work setting up computer networks at the college. This happened despite the department's knowledge of his computing abilities and the fact that his knowledge was frequently relied on by both faculty and management.

Recently, workers have sensed that the Participative Management program has "gone out the window" and that the administration is "basically doing its own thing." Any new developments in the program have apparently been compromised by the new Draft Action Plan prepared by the administration of the college. In this plan, Participative Management has been fundamentally altered since the president gained the ability to unilaterally select members of the Participative Management committee and representation for workers has therefore been discounted.

By almost all accounts, workers agree that Participative Management could be a good thing for them and, specifically, that it could have a very positive effect on learning of all types in the workplace. However, interviewees are equally clear that they have not seen concrete evidence that the program is working. One way that Participative Management principles can be tested, is around the recent invitation for workers to assume more responsibility in employee development. Workers are in a good position to evaluate the Strategic Action plan on Employee Development. The union and membership can question management for further clarification of this section and can develop their own strategic response to the document. The college is calling for higher participation rates in professional development activities by workers and for staff to use the skills that they gain. This study shows that there are a number of barriers that must be addressed before people can become more involved in ongoing effective training: more open access to training, genuinely shared power over decision-making, etc. One question at the centre of it all is how the "relevance" of training is established and by whom? Are resources available and are they adequately distributed in order to support relevant training?

Moving Responsibility to the Worker

In the latter half of the 1990s, on top of everything else, new parameters for staff development, including a "Personal Professional Development Plan" (PPDP) were introduced. This was part of the college's action planning process mandated by the November (1996) Strategic Plan. The college considers each employee responsible for developing the PPDP with his or her supervisor. Guidelines for initiating this process were published in the Centre's winter program guide and there is even a workshop devoted to this topic which falls under the theme of "accountability."

> The PPDP is essentially a tool to enable employees to engage in a self-directed process that guides their life-long learning ... The PPDP process begins with the identification of skill sets needed currently and in the future as identified in the Local Operation Plan of the school or department, job shifts within [the college] and in Society in general. (PPDP Program Guide)

Some financial support for education and training is available through initiatives like the Professional Development Leave Program (PDLP) and Tuition Assistance, but in practice it is quite limited for the majority of unionized workers at the college. The PDLP was not initially designed for union members at the college but for professional or academic staff. At this time, union members had access to a type of apprenticeship program. In the early 1990s the apprenticeship program was cut but the local union successfully negotiated formal member access to the PDLP to upgrade qualifications or to learn a new trade. Still, few union members had actually made use of the program.

The college advocates the benefits of enriching workers' knowledge, both at work and in broader life in terms of its public statements and policies. However, restructuring has meant that many workers have too much work to allow for proper training, or their supervisors do not view it as relevant and/or can not afford to have them absent from their job. Workers would like more information and more choice and control over the work and learning they do. The contradiction arises when, on the one hand, workers are being given more responsibilities than ever before for work-based learning under the banner of employee development and life-long learning. On the other hand, workload has been increased so that there is significant overtime work, both with and without pay, and super-visors cannot afford to give time off for training. In general, the result is a shift in responsibility for training onto the backs of individual workers who are increasingly being encouraged to train on their own time to meet their end of the bargain. In sum, the college restructuring has placed work-ers in a defensive position with respect to their training. The burden of training is effectively falling on individual workers without sufficient sup-port networks or other resources:

> You are asked to do something and you either learn it on your own or you will be put into a position where you cannot do your job. There is profes-sional development that you can take part in, in theory, but with all the downsizing how can you, in all good conscience, leave your co-worker to take a full-day course when they are already over their head. In a way, you can't get your job finished to get trained, and you can't get your job done because you don't have the training. Unless you are willing to de-vote your own personal resources and time into your own education ... there are relatively few opportunities for job training and education. There is also an intimidation process. Competencies are expected and there are deadlines. If you can't meet these, well the approach is not what is wrong here with the system, but what is wrong with the individual that they cannot do these things. (Ellen)

While there was a feeling among workers that the college should sup-port people in their efforts to become more mobile in their careers, inter-viewees also felt that this should not be at the expense of job security. The college has begun to move from a language of "sustained employment" to "employability," and is now talking about retraining for jobs both inside and outside the college. This marks what one worker feels is an "uneasy" shift away from an interest in securing jobs within the college towards dis-placing primary responsibility for job loss onto workers. It is also reflected in the increasing number of professional development courses offered on general themes such as "Trends in the Workplace," "Job Search Strate-gies" and "Retooling Your Resume."

At a broader level, some of the support staff mentioned that employers should bear the burden of training since they are always asking for "expe-rience." Ellen, once again, comments:

> [Training] has to be accessible ... but it should not take time away from family because that puts stress on [us] ... think there should be an element of our jobs that is strongly recommended and flexible in terms of what people want to do career-wise. If that were more a part of your work week, learning and education and/or a job for the future, it would enhance the job.

Many workers felt that changes should be made in job design to accommodate and make use of people's skills. There is also a need for more understanding around the transferability of education, particularly for recent immigrants whose skills are perhaps most underused at the college.

Bumping and Multi-tasking: The Reality of Work and Learning for College Workers Under Restructuring

> I got bumped about three years ago. We lost about fifty people at that time. They call it attrition, normally it's because people retire from the same department. That department's screwed because you can't just spread them out (people retiring) across departments. I'd sometimes rather be laid off than do two people's job for the same pay, you know. Sure the union will tell you "you only do as much as you can for one day" and I'll buy that line. But when you have ten students lined up in front of you, and you know that one's a single mother and that one's blind, you know what I'm saying – they have problems too. You can't just say, "I'm on my lunch I'm going to close the door." (Jackie)

Programs, policy and rhetoric of learning that resulted in a significant downloading of responsibility for workers' learning onto the backs of workers largely without the appropriate support were not the only way to understand public sector restructuring at the college. Beneath them were the reality of insecurity, the threat of outsourcing, the downsizing, the bumping, the multi-tasking, and the work intensification that characterized work and organized learning across the different occupational groups at the college.

Throughout the 1990s the union local has had to deal with lay-offs and it appears to have been relatively successful in advocating the use of retirement packages as a way of lessening the overall effects of the downsizing while management has instituted a hiring freeze. Intermixed with each of these factors have been attempts to reorganize work processes. Together this has resulted in an extraordinary number of workers moving into new jobs (sometimes newly created positions) within the college, sometimes "bumping" workers with lower seniority in the process. Those who lose their jobs during restructuring also have the right to be retrained for other available positions within the college – the Job Stability Fund was created for exactly this purpose. However, the fund came under marked strain because the increasing number of lay-offs resulted in minimal support for worker transitions. Filling the new positions in the context of a hiring freeze has often meant that workloads have greatly increased, while many of the enjoyable aspects of work that staff used to do have been taken away. As in other workplaces examined in our research, college management has been trying to "multi-task" in order to deal with the crunch. That in turn opens up the way for further work reorganization and intensification. Indeed, it is the view of some workers that management's next step will be to eliminate entire work areas in response to restructuring and government austerity. Working in such an environment has produced considerable strain on workers, a topic of much concern for those we interviewed.

The interviews revealed two main reasons why people are, have been, or wish to be involved in training. One reason is to learn skills for their immediate jobs, while another is the need to be "marketable," though again this tended to rely on individuals' own personal resources.

> I engage in education and training to learn something else and make myself more marketable. You really need more training in computers. It is not good enough to have basic knowledge, you have to focus on packages and know them inside out or you are disqualified before you begin. (Carrie)

> I took a computer course at the university when I was at a different campus because I knew I wouldn't get the training I would need to be marketable. So I do a lot of that kind of thing. Night courses. (Alice)

> I don't think I am going to have any chance if I don't get that certificate, or a piece of paper. Unless I am lucky enough that my own institution decides to train me. (Lewis)

Indeed, this notion of marketability, closely associated with the atomizing notion of the "enterprising self" was an expression of the trend to move the responsibility for learning onto the backs of individual workers. While, of course, a certain level of ongoing training is necessary for work in any dynamic organization, a great deal of this learning was organized by the principles of restructuring including this movement of workers from job to job and the prospect of having to enter the external labour market. Office-based work and technicians' work, shipping and receiving make up the majority positions, and as these each involve computers and the ongoing change of software applications, restructuring and training came to be rather synonymous with computer training for most workers. Thus, heightened by restructuring, the need to adapt to new computer technology is a major reason why people are forced to do specific forms of learning for their jobs. (However, most staff members said they had learned something about computers whether they used it in their jobs or not.) Those who have experienced "bumping" at the college say that there is more of a need than ever to have very specific skills that will allow them to move within their pay bands. Learning specific software packages has been particularly important for staff members who have taken over the old clerical and administrative positions, as well as for technicians who now have to learn new communication systems.

> I feel that my line of work is slowly going to be phased out. I think everything is going computer based. So I am striving to learn all I can to learn about that, on my own, so that I can move from one area of the College to another area. So, yeah, I have a real self-interest to try and learn this stuff ... The more I know about this equipment [e.g., multimedia, internet] the better my chances of surviving at the college. (Matt)

Along with the reorganization process and the elaborate bumping and intensification activity, workers consistently reported an inflation of job credentials for new jobs while at the same time many people's existing qualifications, relevant to actually doing the job, continue to go unrecognized by management. The college has made concerted attempts to develop a training infrastructure to respond to some of these concerns but new problems arose and significant failures of these mixed with marginal successes from the perspective of workers were evident. For example, clerical and other office staff at the college have obtained college certificates or degrees that related to their jobs at their initial hiring. However, because of the reorganization of the college, many people have been shifted into jobs that do not necessarily reflect their prior training. Indeed with some notable exceptions which we will address below, most jobs have involved the need to learn new forms of computers.

> I've never been trained to do any of my [several different] jobs at the college. Usually someone has left a job a good two to six months before you come into it because of the hiring freezes and the length of time it takes to get someone in there. Managers don't tend to have a lot of patience with their employees going back to their old jobs to help out. (Maureen)

In the majority of cases based on our interviews, bumping has placed people in new work situations that are difficult to cope with. Several interviewees were actively strategizing about how to leave the current workplace: "When I lost my job in the bumping three years ago, I had no idea that this could happen. I'd made a big decision to come and work here. But now – it's like if something else [a job elsewhere] came along it would be okay" (Norm). Workload stress and hostility from co-workers who feel forced to train new people under already intensified working conditions have made it impossible for some people to adjust to the job options they are faced with.

In general, a college environment poses special problems around learning on the job because every new semester brings changes that affect workers, especially women, given the fact that they form the majority of those working in the administration services. These include learning about changes of procedure, policy, programs and new computer software. Sometimes structural change has meant that almost everyone in a given department is busy learning their jobs from scratch, and this has created difficulties around sharing information and learning freely from co-workers. Even worse, midway through our study management undertook a student survey that suggested dissatisfaction with the availability of student services. The college's response was to extend hours and add shift work that, under the imposed hiring freeze, further intensified work. The contradictions of restructuring on workers' work and learning had reached a troubling level by the end of the study with interviewees commonly stating simply, as we noted earlier, that they cannot finish their work in time to go for training, and cannot do the job quickly because they lack the training.

Restructuring introduces unique meta-learning needs such as learning how to operate within the training system or learning vocabulary/skill to participate with others to learn a new job. Staff members also described how part of their learning involved picking up strategies for "learning how to learn." This was mentioned most frequently by staff in higher-level office support occupations such as coordinating jobs. An older worker said that one of the most important aspects of going to university at an older age was learning strategies for effective studying. Another staff member provided an interesting example of how important it is to have a knowledge of *how* to access learning resources when he described that it took him a while to learn that he could "grab extra books" at the library in order to supplement his technical studies.

Near the end of our study the union organized a meeting on these matters. Discussing the effects of bumping on workers, people revealed that their greatest concerns revolved around the human relations aspect: how people were notified, who was bumped, and were some workers being wrongly bumped into poorer jobs. However, it emerged that the bumping should be the very last resort due to its negative effects on the social networks felt to be essential to the workplace.

In the final instance, the college has consistently positioned itself as the "provider" of training, while staff members' role is to take advantage of the training that has already been developed. Our interviewees were quick to point out that this follows a traditional "expert" model, which assumes that the content and methods of training are best known by external professionals. While the college has formalized some workers as instructors, this has usually been for the purpose of instructing faculty or students at the college, not for staff members' own training. The worker-teacher model could possibly be developed more extensively at the college (without ignoring the problems of job description and wage levels, and the potential conflicts between the staff and faculty job descriptions). The professional development courses appear to be the most accessible forms of staff education, as they are carried out during work time and are generally taught by outside professionals or non-union staff members. They would therefore be a good place to begin developing a worker-as-educator model, as per the practice of many labour unions in terms of their own educational processes. There is formal training available in this area in the college and elsewhere, referred to as "train-the-trainer" and "peer training," but the CED could be asked to provide this training to workers as a first step towards more relevant program development. A beginning step could be to advocate worker-centred learning practices to the providers of professional development.

In general terms, according to the interviewees, several local educational practices need to be developed in order to maximize the learning experience for staff members. Four strategies were suggested. First, the curriculum has to be practical and relevant to immediate situations. Teaching practice needs to accommodate a variety of learning strategies and approaches. Moreover, the curriculum has to be connected to what people are doing, even if it means expanding what they do to match the potential of the curriculum. Second, the learning context has to be sensitive to the audience. Instructors need to have not only an intimate knowledge of their material but the context in which things actually happen, i.e., knowledge based on "having been there." Here the involvement of co-workers is the most direct solution to perceived inadequacies. Third, the content of worker-centred education includes the specific skills needed to do the job well plus a broad knowledge of the social context of work *as well as* the skills needed to advance in the workplace. Fourth, social networking is a preferred mode of learning for many of the workers and this should be positively developed rather than cramped into the narrow confines of a classroom or remoulded to focus on individual, self-directed, computer-based learning tutorials. "Self-direction" should admit both individual and collective direction, the latter of which emerges from social networks which are already the typical bases for much work activity.

Segregation Processes at the College
Internally Segregated Work and Learning

Based on the strategic planning statements, policies and organizational structures in place, the college would seem like a learning oasis, but there is distance between rhetoric and reality of workers. A close inspection of the changes reveals several gaps. Here in this final section, however, we explore how learning in a knowledge society is also unevenly distributed *between* specific groups of college workers.

Resistance to workers' desire to learn comes in the form of access to professional development where it is difficult to find other people to cover the job. This problem disporportionately affects workers who have the least organizational authority and discretionary space. Moreover, difficulties emerge where there are meetings that interfere with course times.

> There are a lot of courses I'd like to take, but you take time off during the day and there's no one to cover for me. It's very hard. I had an interview with one of the people from the professional development department and they said they'd try and offer more in the summer, but that's when I take my holidays. (Cathy)

Inflexibility to accommodate women workers who have primary home or childcare responsibilities also means certain workers are left out. Those without immediate support systems (like available friends or partner, and/or extended family) have particularly limited time. One staff member discusses the impact of work schedules on family time and on access to courses:

> I'm still thinking about taking a computer course, but with the little one ... and my husband working every other weekend, it's difficult ... You're working to pay the bills. I look after my daughter in the day, my husband comes home and looks after her when I go to work. I don't see my husband except every other weekend. (Lilly)

With the shift of the responsibility for training onto the backs of workers on their own time, a significant numbers of interviewees experienced difficulties in computer learning because they had no access to the appropriate level of technology in their home or community. In some cases courses are simply full but it is also true that there is little recognition of barriers to participation facing workers while "priority access" to courses is not extended to those who are most vulnerable. In still other instances, workers attend courses or learn informally with others, but see skills and knowledge capacity slip away from lack of opportunity to use them in the workplace.

> You are just expected to know your job here. There is also a lack of time to keep ourselves current. They buy data bases and you are expected to learn them yourself – they say just play with them but you have to have the time and it has to be connected to your job. If you learn the computer and then don't have the opportunity to use it for months, what good was it? You will forget. (Lacy)

The problem of resources was also linked to increased administrative and managerial duties that support staff are required to assume. There is a feeling that while support staff are being given extra responsibility they are not being given the means and recognition for carrying out these new responsibilities:

> As an office support person ... I'm doing things that faculty would be doing, or administrators. But, I've always had to kind of fight to get the necessary resources to do things and the permission to do things. So now, I just go and do them. Because there's a perception within the bureaucracy of the college, that certain things sustain, for instance an administrator's position, they have to be doing certain things. And so, a support staff person shouldn't be doing that. (Connie)

Several of these issues return us to the importance of authority. While this plan for employee development (i.e., the PDP) emphasizes continual learning it is closely tied to the amount of resources negotiated at the individual level. Both of these issues set the stage for uneven access to resources and a restricted path of professional development for employees. As noted in the previous section, there is a great deal of variation in how PDPs were viewed by workers, with some finding them helpful to their jobs and others finding them untimely and irrelevant. Getting time and permission to attend professional development courses, and the relevance of courses to both the current job and a vague notion of career progression, were viewed as keys to participation in professional development courses. Permission seems to hinge on a notion of "need" that has been set arbitrarily by the college and mediated only by perceived importance, authority and discretion linked with position. There are no stated rules around when, how and for what kinds of training one can get permission. For example, some support staff reported that management provided no in-house courses and granted little, if any, leave to take courses elsewhere.

In addition to the regular professional development courses, paid educational leave is a way for support staff to develop professionally. However, the organizational policy indicates that paid leave is mainly used by "professionals" or people with a "need" to continue their education at the advanced level. Many workers seemed to have bought into the biases inherent in claims that professional work is inherently more complex and this is perhaps a type of cultural barrier for the union to address. It became clear in the interviews that the idea of "professional development" did not resonate as something relevant for all workers. It is exclusionary both in terms of people's popular conceptions and in terms of informal policy because not all work categories are considered professions. But, at a more basic level, workers simply are not aware of paid educational leave or how they would access or use such an education strategy. In order for some staff members to take their professional development seriously, they would need to see their work linked developmentally to other training and job opportunities at the college.

Job Segregation and Training Resources

Access to training resources can be linked to the issue of job segregation at the college.

> A lot of staff, clerks, took the adult education courses … but the problem is, not a lot of staff have progressive managers who are going to tell them about it. Especially areas like registration, those people are very busy and they don't want them taking two hours to learn something that's not related to their job. (Jackie)

Of all the occupational categories, administrative/coordinators and technicians seemed to have the most flexibility to access resources for training. Authority even aided these groups when permission was denied for external training, as when, in response, coordinators organized to have the course offered for students at the college and then promptly enrolled themselves. Part-time staff and security workers are the least likely to participate in professional development. Not only are they not notified consistently of training opportunities but many custodial and security staff often work night-shift, minimizing the availability of courses.

There is evidence that custodial workers and, in particular, housekeeping workers are, in practice, systematically excluded from staff development. At least part of the problem is that courses do not address the issue of ESL among these workers, but also the courses offered do not appear to be significantly related to the ongoing career development as the job is treated as a dead-end, terminal position. In other words, when the jobs do not offer new opportunities for work, it is more difficult for people to justify training needs – especially when training opportunities have to be relevant to work as stated in the strategic plan. The need for broader employment possibilities in the custodial area is expressed thus: "In custodial work, we should be able to do things like paint and [revitalize] the classrooms, not just wash floors" (Ken).

Although the college was a forerunner in Ontario implementing employment equity, and more recently pay equity, and the college hires 50 percent men and women, there are still considerable problems linked to gender segregation of work. This occurs between and within job categories. For example, both men and women do custodial work but women primarily do housekeeping (i.e., hand cleaning), whereas the men (e.g., custodial lead hands) operate equipment (e.g., buffing machines) and are therefore in a higher pay category. Female housekeeping staff who do not read and write in English are doubly blocked from becoming lead hands, because lead hand work requires written English skills. Another gender issue centred on control over workload and more specifically discretionary control over work with women largely doing jobs requiring immediate attention: "We women are all in the jobs where you have to put out fires, whereas the men [lead hands and maintenance] are all in jobs where there's waiting lists" (Lilly). More broadly, because some staff members do not speak or write English fluently, they are excluded from courses that might interest them. Through a denial of relevant resources which legitimately align with the college's stated position on advancing the career opportunities of all employees, English-as-a-Second Language (ESL) workers are excluded from job advancement within the college. "We have people here that don't write or speak English. They can take an English course. But they're not pushed by the employer, like 'Here, take a course, we'll pay the $10 to learn how to read and write English.' Which is sad" (Ken). Furthermore,

> In custodial work, you can't become a lead hand unless you can read and write English which leaves out a lot of very intelligent people but they come from another country and they're older and just don't have the written. They can speak English ... but can't write. They have to do shipping and receiving forms. There must be something we can do ... it's a barrier. (Jackie)

There is a need for English to be made available to some of the workers to enable them to deal with written materials relevant to their work areas. As in other workplaces, the need *is* responded to by workers in this situation despite the lack of support, as people collectively band together to share scarce resources and expertise. For example, many workers with limited experience in speaking, reading or writing English get together with co-workers to help them with these areas. Although some workers may be resistant to, or unable to, put more of their time into learning written and/or spoken English, it was widely felt that the college could make courses and information more accessible and provide encouragement to workers who may want to take advantage of it. English-language skills are becoming more necessary in the context of downsizing where job stability depends on having them, and the movement of workers to different jobs forces them to learn sometimes without an established social network.

> In housekeeping and maintenance we have a situation where at one point they didn't need English to do this job, but now things are so computerized that it is suddenly so much more important for them to have computer skills and read and write fluently. Now they can get bumped into jobs that require not only reading and writing, but fluency. They may not have grade eight or computers. They are working in an educational institution and they are illiterate in English and I do not understand how that is possible in this college community where we are supposedly committed to professional development and education. (Frank)

It is a bitter irony among these workers that the college has all the tools at its disposal for teaching such skills and considers its English Language Institute (which serves international students entering post-secondary education in Ontario) as one of the most advanced multimedia language learning facility in Canada. Yet, workers are consistently left out in the cold.

Comparing Learning Across Groups

In general terms, faculty and management receive the lion's share of training resources on a per person basis. Training amongst unionized workers, as we saw in Table 5.1, is also unevenly distributed. Figure 5.1 shows the relative differences in the reporting of learning resources between occupational groups. It is based on ratings by different occupational groups on how useful various (informal and formalized) educational resources were to their work. It emphasizes a number of things including the overall range of learning resources that workers make use of at work and the importance of informal learning and incidental contact with co-workers and even suppliers. It also suggests some of the differences between different occupational groups with those with the greatest authority and discretionary control generating learning that is most applicable to the workplace. Of course, this does not mean that these groups are doing the most learning for work, but simply that what they learn, according to them, is integrated and applicable to the work process (as defined by management). Co-workers, union training, conversations with suppliers and job related courses were, overall, the most highly rated forms of educational resources.

Figure 5.1: Patterns of Learning Resource Use Between Occupational Groups (college)

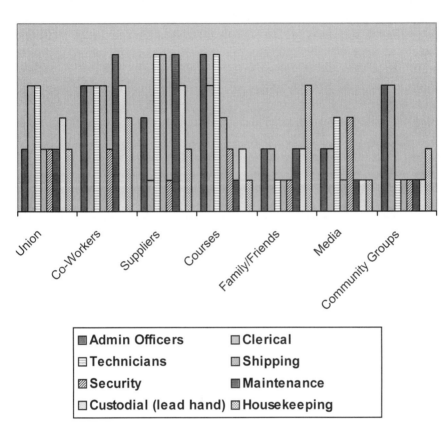

More specifically, we are addressing the idea of a knowledge/power theory of work-based learning. Informal learning, in its general form, is more or less uniform across all occupational groups: people learn all the time irrespective of their social standpoint. Distinctions can be made within this learning, however, by applying notions of authority, legitimacy, power, resources, etc. In the paid workplace, there are significant differences in the nature and integration of this informal learning. The need for computer knowledge is one key issue around staff training at the college. This kind of training is least available in areas like housekeeping, or shipping, where there has been relatively little technological change due to computers, while the most computer training is available in the computing services department. According to Lewis, "[t]here is a lot of training here [in computing services], mostly off campus, or someone comes in." Workers in administrator/coordinator and clerical positions are generally left to their own devices to learn new software or to rely on professional development workshops. Office workers in the continuing education department have been in a different position from other staff because they have access to free courses offered in their department. Technicians and trades/maintenance workers as well as coordinators and program officers have the highest levels of discretionary control with their informal learning tightly integrated with their work. Clerical, security, male custodial and shipping workers have medium levels of discretion in their work and integration of their informal learning. Female custodial workers experience the least discretion and least opportunity for integration of their learning into their work (see Figure 5.2).

Figure 5.2: Current Relationship Between Job Power/Authority and Integration/Support for Work-based Learning

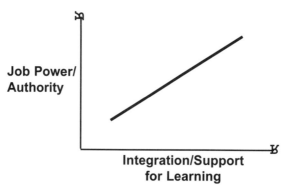

This produces a very basic relationship by which most, if not all, training interventions (in response to restructuring or any other impetus) can be judged. The relationship represented in the figure links to the fundamental issue of how well new forms of labour process mobilize the hidden learning capacities of workers.

From our interviews, it appears that access to formal job training is linked mainly to one's job classification (which is related to prior training) and to the department one is in. In addition, although most job categories have experienced increased workloads, there is some evidence that women in administrative support services bear an undue burden of work. At a more general level, access to training is linked to the amount of influence people have in their particular departments. Higher-level coordinators, who have relatively more flexibility in terms of time, have more opportunities than other workers to learn broadly for their jobs. People who work alone at satellite campuses appear to have the least opportunity to take advantage of training offered during work time. Two other factors seem to relate to whether or not people had taken advantage of any sort of formal training program. These included family care responsibilities and English-language experience. Several participants stated that workers with no experience of written English, primarily those in the custodial areas, would have been excluded from most of the training available at the college.

Concluding Remarks

This chapter has explored the effects of restructuring in a public sector college on work and learning patterns across different occupational groups. This site analysis reveals an example of the use of progressive managerial rhetoric that, when investigated more closely from the standpoint of workers, is shown to be deficient. There are significant gaps between policy and practice. Nevertheless, there is a basic training infrastructure at the college with a great deal of potential for workers if truly participative forms of working and learning can be established. Through this, better decisions regarding work intensification, multi-tasking etc. can be implemented, and innovative, worker-centred responses to public sector restructuring might emerge.

A key finding of the chapter is how race and gender play a role in shaping work and learning opportunities, when mediated by the occupational categories. As we climb the occupational hierarchy, more training opportunities and, equally important, more opportunities to apply learning in all its forms tends to emerge. A more equitable distribution of opportunity would go a long way to achieving the kinds of results in which many members of the union are interested.

CHAPTER 6

Divisions of Labour / Divisions of Learning in a Small Parts Manufacturer

Anne Jawarski works as a product inspector at the parts manufacturing firm. Like many of our interviews, Anne's interview took place at a spot close to her home, a local donut shop near her apartment. But the experience was often difficult for women like Anne. In fact, according to Anne, being so close to home made it more challenging to focus on the questions and we could sense her difficulty in leaving aside the work that was waiting at home along side her two boys and partner.

In her late 30s, Anne says she's "one pay cheque away from poverty" with two income earners "barely making it ... that's my reality." Growing up didn't afford Anne much of an education. Poverty, the need to bring in money to the home and isolation all made attending school difficult and attending anything beyond high school a vague fantasy: "Education? I would have loved to have gone but wasn't able ... I didn't have a clue about how to go about it past high school and couldn't afford to anyway." Upon graduating high school she faced a recession and difficult prospects for finding a job. She says she literally cried and begged for her first job working as a clerk at a nearby department store, working several years there before arriving at the parts factory at age twenty five. Now she describes herself as a "little sponge," learning both formally and informally in whatever periods and spaces she can find: on the bus into work, by the light of the bathroom once the children are asleep and dishes done, on breaks at work, with truckers dropping off materials in the shipping yard, in the change room with co-workers at the end of a shift, and on and on.

Becoming a product inspector at the parts and parts factory; a position she finally gained after having been turned down for promotion several times before, is a real change, she says, from life on the production line. She had to endure a long series of training courses to get this job, made even more difficult against the backdrop of her double day (responsible for both bringing in an income and taking care of the housework) and the bitter fight she and her union has had to wage against the company who wanted to hire an outside (male) applicant who, in fact, had both fewer formal qualification and no experience working in the factory itself. Indeed, it's a subject that Anne bristles at as she sips at her coffee, and with this the conversation turns to the many examples of the divisions at work; between management and workers, between men and women; between trades and production workers; and between different ethnic groups. It's a workplace where people are frightened. Workers feel pride in their work, and according to Anne have a lot of ideas about how to improve production, but they still feel insecure. Women, in particular, get laid-off frequently, and there's constant talk by management about foreign competition and the need for everyone to simply work harder if they are to survive.

Anne is Canadian born and in this way her profile doesn't match the majority of workers at the parts factory who either grew up speaking a different language or, more often, came from another country altogether. But her struggles as a working-class woman help to introduce a general dynamic that helps put a human face on the types of challenges facing the majority of workers in small parts, light manufacturing. Her movement from the production line into a more skilled position as a quality inspector is also instructive because it sheds light on experience of an internally segmented labour market where workers of colour, women workers and non-English speaking "ethnics," as white workers at the factory call them, face difficult barriers in terms of both work and learning.

Introduction

Our research in the small parts manufacturing sector involved a small auto parts manufacturing plant that was a supplier for one of the large North American auto assemblers. The company was established in the early 1960s and operates four facilities: three in southern Ontario and the other just inside the U.S. border with a total workforce, across all plants, of approximately 450 people. The plant we focused on was the oldest of the four and operated using well-established roll forming, stamping, extrusion, co-extrusion, injection moulding, stretch bending, anodizing and basic assembly technologies which, even in the early 1990s, were considerably dated. At this plant there were just over 100 unionized workers, although during our study this number fluctuated by as much as 30 percent, workers being on a cycle of lay-off/call-back according to workplace orders. Shortly after the research ended this plant was closed and the staff merged at a slightly larger, though similar, facility in the same city.

The majority of workers at the plant are women (approximately 80 percent) who work on a combination of small continuous run production lines and small batch processing clusters. Most women work in these production positions. A small number of unionized female workers are quality inspectors while male workers are mainly skilled tradesmen and set-up/machine operators. With only two exceptions, all of the technicians and skilled trades workers were men. The workplace encompassed a wide variety of ethnic cultures. Many of the workers across all occupational groups are recent first-generation Canadians immigrants from Italy and Eastern Europe, but there are also workers who have immigrated from Asia and South America.

In terms of the changing conditions of the sector and its effect on the specific character of the firm, besides the reconfiguration of facilities, the company was approaching a fundamental turning point. It was a small, privately owned automotive trim manufacturer, attempting to meet the demands of an increasingly global market. Previously the firm was positioned geographically to take advantage of the U.S./Canadian trade policy known as the "Autopact" which guaranteed that the percentage of vehicles sold in Canada had to be made in Canada. As of the late 1990s, following the establishment of the North American Free Trade Agreement (NAFTA) in 1993, about half of the company's sales became destined for other-than Canadian markets in Latin America, Mexico and the United States. The plant was only able to (cost-effectively) run a single production shift and came to rely on an extremely flexible labour force who became subjected to lay-off/recall cycles that saw its workforce vacillate. In the four years of the study alone the workforce experienced two lay-off/recall cycles. As such, it relied on laid-off workers who, although obviously preferring more stable conditions, remain responsive to call-backs, as well as other contingent workers who appear on a "just-in-time" basis.

Nevertheless, the company continued to survive while combating competitive pressures and changes in the sector with attempts at quality assurance programs and managerial training initiatives centred around "Team building," "Quality Production," and movement towards a "Learning Organization."[1] In these initiatives, however, it becomes clear that the development and integration of production workers' knowledge, despite the formal tenets of the "Learning Organization," were not really part of the program. In the Introduction we characterized this as an "adaptive" rather than a truly "innovative" managerial strategy (Lazonick, 1991) since the specific labour process, including technologies and control structures, continued to match traditional scientific management principles and practice. Closely related to these strategies was the approach to training and learning (again see Introduction) which was not oriented towards the inclusion of worker skill and knowledge creation as a key component of how the firm operated and developed. From the perspective of production workers, claims by the Auto Parts Manufacturing Association that investment in training amongst auto parts manufacturers has doubled in recent years[2] should be approached with caution.

The broader sectoral context gives a better picture of the company's current situation. On the surface, the automotive parts and accessories sector was in a healthy state of employment growth with employment having virtually doubled between 1980 and 1996. Between 1989 and 1992, the cumulative growth rate in the sector was relatively weak, and even declined at some points. Since 1992, however, the industry has shown strong growth lasting into the year 2000. Over a period of fifteen years, the cumulative growth rate in the motor vehicle parts industry in Canada as a whole has almost doubled. The automotive parts sector has absorbed a good deal of the employment out-sourced from large auto-assembly plants into the sub-sector. And, along with this, one of the most important factors contributing to the "turning point" faced by the firm is the fact that there is currently a consolidation process occurring in the sub-sector. Specifically, due to out-sourcing, greater, capital-intensive operations are increasingly required by firms as they assume greater responsibility for product development, higher quality assurance and, most significantly, according to Industry Canada reports (1997b), full warranty responsibility for parts or components. All of this makes the survival of relatively small firms difficult.

These additional requirements favour larger firms which can more easily accommodate shorter production runs, "Just-in-Time" manufacturing and other pressures faced in the new sectoral environment. Indeed, since the mid-1980s the number of Tier 1[3] suppliers to auto assemblers (this company is a small, Canadian-owned, Tier 2 supplier) has contracted radically from approximately 2,500 in 1985 to a projected 375 in 2005 with the domination of firms including Magna, the Woodbridge Group, A.G. Simpson and ABC Group, to mention only the Canadian-based, multinationals. According to Human Resource Development Canada (HRDC, 1999), the cost pressures are increasingly passed on in turn to the smaller

Tier 2 companies. Canadian ownership is concentrated in the smaller Tier 2 suppliers despite Canadian-owned giants like Magna and Woodbridge.

Figure 6.1: Canadian Auto Parts Employment

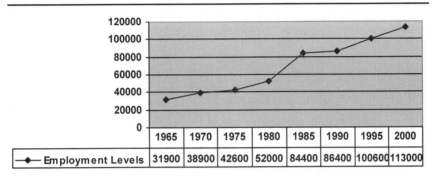

	1965	1970	1975	1980	1985	1990	1995	2000
—◆— Employment Levels	31900	38900	42600	52000	84400	86400	100600	113000

Source: Industry Canada "Canada's Automotive Industry 2001"; www.strategis.ic.gc.ca

Over 40 percent of parts firms have less than fifty employees while over three-quarters have less than 200 employees, but when compared to other manufacturing sectors, the percentage of companies with more than 200 employees is quite high (twenty-two times higher) (HRDC, 1999). Research and development spending throughout the sector is significantly below the Canadian average for manufacturing as a whole and spending fell in the last half of the 1990s. This trend is particularly problematic for small Canadian-owned, Tier 2 suppliers, like the study's firm, who cannot inherit technologies from larger foreign-owned companies, do not have large investment capital to draw on and who, subsequently, have difficulty modernizing and engaging in "innovative" (versus) "adaptive" strategies.

Trade agreements such as NAFTA also produce a selective pressure and demand either large, capital-intensive firms or specialized niche production firms. According to industry profiles (e.g., HRDC, 1999) trade agreements such as NAFTA open the door to capital intensification through corporate mergers in distribution, marketing and rebuilding stream of auto parts sector as well.

These sectoral and firm characteristics play an important role in shaping work and learning practice at the parts factory. Despite the apparent good health of the (sub-)sector as a whole, difficult challenges lie ahead. As a direct response there appears to be a growing recognition by the company that workers' knowledge is increasingly important for the company's survival, although the rhetoric largely outweighs actual practice. By the end of our research at the site, the company was in the midst of a recommitment to the "Team" concept and continuous improvement as part of their quality (QS and ISO) strategy. Through new "Brainstorming" meetings (i.e., quality circles), the company was hoping to make better use of workers' skill and knowledge production[4], all against a backdrop of threats of decreased sales orders, the loss of major contracts and so forth, parlayed to workers through company newsletters.

Nevertheless, this new labour relations strategy can be seen in the telltale shift in company language ("employees" are now called "associates" in company literature), as well as in recent management courses[5] investigating the implementation of the "Learning Organization" model. The company's willingness to engage in discussions of further worker training is another sign of such a shift. QS and ISO certification/audits (every spring) also provide workers with the opportunity to negotiate some of their training/working conditions with the company because workers' cooperation is important for successful evaluations.

Research Overview and Participant Profile

In total, over forty hours of in-depth interviews and nearly 300 typed pages of interview notes and reports were analyzed in order to reach the conclusions of this chapter. We sampled workers from the five main occupational groups on the shop floor and attempted to represent a cross-section of the workforce at the plant based on gender, age, job-type, formal education level, seniority and union activism. Fifteen workers participated in the research (approximately 15 percent of the workforce).

Maintenance/skilled trades workers carried out the standard range of activities associated with factory maintenance including repair and installation of machinery and upkeep of facilities. These positions require trades tickets or, as in the case of one interviewee, equivalent experience, and involve both unexpected and scheduled repair/maintenance of equipment and plant. These positions are typically less routine and involve a large degree of self-direction and the workers had access to their own area called the "tool room." The maintenance and trades, as reflected in the interview sample, are all men. Quality Inspectors were workers responsible for Statistical Process Control of output. This involves quality inspections of parts on the floor and in receiving, and also involves some self-directed work. The position requires the completion of in-house informal training and the completion of 50 percent or more of a Community College Quality Assurance (QA) Certificate program. The machine set-up was carried out by semi-skilled, lead-hands who changed settings on various machines associated with continuous line and batch processing based on the specific product run. This work required the changing and adjustment of production machines including dies. The position offers some degree of self-direction and requires experience generally gained within the plant as a production line worker and/or the "Set-up Trainee" position. Material-handlers, with the exception of one woman (not interviewed), were male workers with previous experience as production operators. As material-handlers they were required to deliver materials to the line, clean up materials, cover for production line workers, etc. This was generally understood to be one of two entry-level positions with no formal skills qualification, but in practice these positions were occupied by men with previous production experience at the plant or elsewhere. The material-handlers experienced some degree of autonomy and self-planning. Production operators worked on the continuous assembly line or in the batch processing groups. The production operator position has no formal skill requirement, and demands the attention of workers virtually all the time the line is in operation (even if some workers were involved in batch work), thus limiting autonomy and discretionary control over work. As reflected in our sample, production operators are predominantly women.

In terms of formal schooling amongst all the workers interviewed, it ranged from those who had not completed elementary schooling (from several different countries) to those with recognized technical training such as that of a millwright or tool-and-die maker, and some who started but did not complete a post-secondary, academic program. We sampled a range of seniority levels in the factory, between one year and thirty-two years. Finally, having noted that a variety of non-English languages were spoken on the shop floor we were careful to sample a range of workers with different first languages, many of whom only had basic English.

Profile of Learning in a Small Parts Plant

Perspectives on Schooling and Work

The work context in this research has some similarities to the garment site in the next Chapter 7 in that it involved, relative to today's standards, low-tech production and what we called adaptive industry strategy with minimal computerized technology. As in Chapter 5, the workforce was ethnically and linguistically diverse. Workers were frequently employed and laid-off and, with some exceptions among certain occupational groups such as the trades or quality inspectors, most had low levels of schooling and limited opportunities within the workplace to engage in discretionary learning. Factors such as these built on past experiences with, and perspectives on, schooling. In this section, we see how the interviewees talked and felt about formal schooling and its relationship to work. One important similarity between these and other workers discussed in this book is that they generally felt that experience (that is, informal learning experience) was the best teacher. Nevertheless, workers' past schooling experiences also had an important effect on them. It shaped how people think and feel about schooling obviously, but it also affected people's feelings and perspective on non-formal courses and other forms of organized training (including those put on by the union or company). As mentioned briefly, the workers had a wide range of formal credentials and several had experienced military training in another country. Formal schooling was obtained in a range of countries including Argentina, Ecuador, Philippines, Italy, Eastern Europe as well as various parts of Canada. People immigrated to Canada at different ages, some coming before school age, others arriving as old as thirty. The experience of knowing another place and culture beyond Canada clearly gave people valuable insights on the plurality of perspectives and approaches to learning. One maintenance worker put it this way:

> Everybody has different experience. I told you, maybe, you make this pencil, use it one way. I use different way. I have to see which way the best. Is it my way or your way ... A lot of people have different experience, especially when I work in another country. We have people from maybe eight or ten countries: Italians, Spanish, Germans, French. We have Croatians. We have Hungarians. We have Portuguese. Everybody have different idea for doing the job.... Normally, it's one way, but a lot of people have different ways to make things a little better, or faster. (Carlos)

Indeed, this openness to seeing things in different ways and doing things differently was an important learning resource at the plant that seemed to aid connections among workers despite the language barriers. Despite coming from different parts of the world, people also noted similarities in their formal schooling experiences.

For the majority of production workers, it became clear that general schooling was not as important as any applied studies they carried out. General schooling was discussed with a mixture of feelings and memories. Despite the wide-ranging backgrounds in schooling experience, it was common for workers, and in particular but not exclusively for those from outside Canada, to recount strict discipline of schooling and its negative effects.

> I learned discipline in school. We had this nun that punished you with a stick for poor grades.... You had to try hard to learn or she would call you to the front and give it to you. We were afraid. I got it a couple of times. (Ricardo, born in Columbia)

> I liked school in lower grades until grade 7 when I had a miserable teacher. He would throw things at students, almost got fired when he

smacked a girl's head between two books for talking. That's when everything turned around and I started losing interest. (Jodi, born in Canada)

Beyond the theme of harsh discipline, many talked about schooling as a waste of time in connection with their adult lives generally and their working lives at the factory specifically. Many were eager to get out into the world of work and some had to leave school before they wanted in order to help support their families. A minority of workers wanted to pursue their formal schooling but did not have the resources to do so.

I went to grade eight … I couldn't go to do more school because we had hard times. I was lucky to go to grade eight. (Mary)

It was so hard to find a job growing up … I was trying to attend school and to get into university, but we didn't have enough money to do it. (Ricardo)

The difference between formal schooling and learning was an idea that resonated through many interviews. Most interviewees said that schooling was a difficult experience, while learning (sometimes within school but mostly outside of it) was something they did voluntarily, with pleasure. One worker, Rosa, refuses to be categorized according to her level of formal schooling. More than this, she talks, as many interviewees did, about how she overcame skill deficits collectively. Rosa was helped by family members, others are helped by co-workers. The difference between schooling and learning for these workers revolved around the option of drawing on the help of others openly when not in schools. Drawing on others was not a source of embarrassment, but was seen as a valuable and legitimate resource for people's learning.

Interviewer: I'm going to ask you a little bit about your schooling background.
Rosa: School? Not much there. I don't care. I'm not ashamed. You know, I can read, write in my language. Only here, I can understand a little bit.…
I do everything by myself. Only some things my daughter does for me, to write or read in English. But I have no problem. I did good for myself. I'm married with kids. I've worked to buy two houses for my kids. I have a beautiful house for myself. I worked hard. I did good for myself.

When discussing the need for formal schooling specifically on the job, even those with highly technical training agreed it was opportunities such as "seeing the whole process" (Vlad) that mattered in terms of both doing and learning about the job. There were also a number of different issues woven through these discussions. The value of learning and schooling shifted when one discussed "getting the job" as opposed to "doing the job."

> You need basic math, some high school ... about grade 10 is needed for the job. (Marco)

> I don't need any formal qualifications to do the job. I haven't used college courses on the job. I wouldn't even need high school to do the job. Maybe grade 10 and English skills. Grade 12 would be better, for statistics. Some things are computerized. (Anne)

> More than 50 percent of it [technical education] I don't even use here. I found it really interesting but I don't use it. Didn't need high school, if you've worked here for a while, if you know the parts and stuff you could do the job. I've learned so much more from doing the job than I ever did from going to college. If you've been working the job and you've got the experience with everything in there, that knowledge in itself saves so much training time. [But] it's like they don't like to promote from within ... [so] it's difficult to build your skills. (Jodi)

It seems that workers needed only the basic skills offered by formal schooling. Several workers without even grade 10 schooling operate very effectively in the plant and did not seem to benefit from formal training. "I have to know about computers because the machinery has changed now because of some computer-driven presses. But you just learn on your own as you go" (Daniel).

However, while formal schooling was not viewed as central to the ability to learn things, workers recognized that it could have an effect on one's feeling of competency and status, as well as job mobility and job prospects. Schooling was sometimes said to be a place where personal development skills could be learned, including "communication," "patience," as well as foundation skills like basic math and language (especially English).

> I learned patience from my education ... to get along with people. It helps me a lot to communicate with my co-workers. The education level of the people at work is the right amount because if they had high levels then they shouldn't be there. They can get a better job ... If I went to school, I would expect a different type of job. (Ricardo)

> I would never cut back on anyone who wanted to get an education. I would have loved to have got one but I wasn't able. It would be terrible to stop anyone. Any form of continuing education is always good. It makes a person feel better about themselves, more qualified, more options. The re-organization in the workplace leads people to feel far worse because they have the skills and training and they're not able to use it. You can never have too much education though.... In a lot of jobs, the requirements are ridiculous. I'm not talking about surgeons, but your average working people ... most of it could be on the job training. If you make the effort, it doesn't take long to learn. (Anne)

In other words, the opportunity to gain experience, knowledge and skills is the key to workers' learning in this parts manufacturing plant. However, the opportunity to learn new things in the workplace is not universal. Rather, learning opportunity was related to specific job categories, with some job categories giving more freedom to move around in the workplace, to see new things and to communicate with others.

Anne's comments begin to explain what some workers may mean by "smart," "able" or "intelligent." Again, this is closely tied to opportunities at work. The workers in this study almost universally refused to limit intelligence or "brains" to natural talent or even school credentials. Instead, they preferred words like "ability" which was not based on individual talent but was a product of experience and opportunities to practice.

> I don't call it smartness, I call it ability. If you go to study and learn it successfully then you have gained ability. If you go to a job and you can fix things then you have ability in that area. It's all "ability." (Ricardo)

> They say intelligence goes down after 35.... I don't have that feeling. If my intelligence is going down, and my experience is going up then I don't have that feeling. If you walk into the shop and you begin to find out what is what, that's a kind of intelligence. I just don't think education and intelligence go together. (Vlad)

> You generally don't need too much education. You basically need to give people a chance. (Mary)

This distinction between formal schooling and general learning opportunities is important. First, people indicated that the level of schooling is often used wrongly in society, as a measure of success and even a person's worth. When workers talked about learning based on opportunities for experience, they seemed to critique a school-based credential system that, to them, appeared to be arbitrary. However, entry-level requirements for many jobs (across various workplace contexts) are becoming inflated. This puts a squeeze on people's employment mobility, especially for those who have not reached a grade 12 level in the Canadian system, or those whose trade certifications are not recognized in Canada.

Access to Training at Work

The parts company provided some training for their employees in production areas. This included basic training (e.g., Punch Press Safety, Fork Lift Licensing), health and safety (e.g., Workplace Hazardous Material Information System [WHMIS]) and quality control (e.g., ISO-9000, Quality Assessment). But these are all mandated, either legally by the provincial government or by quality certification standards like ISO. The company also uses "trainee positions" as a form of training. Set-up, extruder set-up and operation, and maintenance positions all have "trainee" versions through which a worker can learn on a one-to-one basis from a more experienced worker. However, the opportunity to take on a "trainee" position is unevenly distributed across workers. Access seems to depend on position as well as gender:

> *Interviewer*: So basically in terms of an education program in the workplace, there hasn't really been anything like that?
> *Mary*: No, the company updates us, because we use so many kinds of oil. There are opportunities through the union for us. Time to spend one day with somebody to explain us, what they use in there. *[Recycling]* all the water, whatever we use in the plant. Like now *[with WHMIS courses]* they want a label attached on bottles, because if I pick up this bottle I have to know what kind of stuff is in there and if I use it, does it cause damage on my skin. They send us for one day, like a school. They bring a teacher. They explain us everything about that. *[And]* every three months we have a safety meeting.

> They don't have any courses [for people] where I work. They have them for other people. Quality people go out. They say they have lots of courses now [but] not really. We don't get them that much, the people who work on the line, people like me. They don't let you know so many things, you know. (Rosa)

Some workers reported having taken no company courses, while others, including quality inspectors or set-up workers, reported having taken as many as three courses in the last five-year period. Trades and maintenance workers typically require their trades papers, there is no "trainee" position designated in the collective agreement, and there are no training courses related specifically to this work (these workers do participate in some of the general course offers, e.g., WHIMIS). However, at least one maintenance man (Lew), years earlier, had moved into his position based on informal learning as a material handler and then set-up person. In terms of training, though, the average number of courses per interviewee over the last five-year period at the plant was less than one course per worker (0.6). Workers reported that they had spent an average of 7.7 hours of in-class time (ranging from zero hours to as many as fifty hours).

Participation in company courses primarily appeared to involve the inspectors and set-up workers, rarely the production operators who comprised the majority of the workforce. In general, those with the highest educational credentials tended to spend the most time in company courses and other schooling, which emphasizes the relation between power and knowledge production. The trades and maintenance workers participated in some company courses at about the same level as those with lower formal schooling credentials (generally the production workers). As we note elsewhere, they have the most free time and movement within the plant, allowing for more discretion and more informal learning in the workplace.

There were also other patterns of access, involving gender, age, and level of schooling, to courses visible in the plant. First, younger workers tend to have higher formal credentials than older workers. This made younger workers more comfortable in classroom settings and, in the cases in which prerequisites are required for advanced courses (e.g., quality control, machine set-up, hazardous materials), made courses more accessible. A second, less immediately apparent, pattern that will be discussed later in the chapter is the gendered differences and the linkage to formal credentials versus informal learning experiences and promotion.

There are differences in how workers viewed training opportunities provided by the company. Some workers, such as Anne, are generally critical of their experience with the company and their training.

> They look at people as uneducated labourers. To a certain extent, I'm not very trusting, [because] I know what courses and non-courses my bosses have. I don't get taken seriously. I don't think anyone in the workplace does ... They look at everybody as being irresponsible. (Anne)

She goes on to say that sometimes company courses have few real effects on the work process and are more about show and sales.

> Employers supported worker education in our department because it makes themselves look better to their customer. You have people that are qualified ... There were no surprises [with respect to any courses that were offered]. The daily activities don't change. Just some minor changes and adjustments ... I was thinking a lot of learning could be done with the company's help ... Benchmarking. People training, cross-training, learning seminars, available at lunch, available after work. It doesn't necessarily have to be on company time ... Most companies won't do it unless it looks good for them for another reason, so they can say 'my employees are skilled.'

Furthermore, skills training without an opportunity to *use* the skills, is not only frustrating to the worker who invests time and energy but even bad for the company:

> A lot of Canadian companies are still back in early 1900s, the CEOs, the managers. You know they talk about people empowerment. It could work [but] they don't want to teach us too much, just enough so we know what's

going on ... A lot of our bosses don't have as much education as [some of us] have, so you don't want your people to have any more than you have. You need to feel safe. They [the company] don't want trouble. (Anne)

As with most industrial training that companies facilitate, the bulk of training often takes place "on-the-job." As a component of a company's training program these efforts are usually viewed positively by workers – probably because in general terms this "curriculum" is taught by co-workers who have seen processes from the workers' own point of view.

The company for me has been very great because they give me all the opportunities [to train] ... I probably came up with the right answers though. (Lew)

Many of them might not be interested in doing any more than what they're doing. Maybe some of the younger people or the guys. They try to have safety meetings. I think they should have tours into different parts of the plant so that you can be more aware of what you're actually doing. That would be a boost to everybody's morale. They'd be educated better as to what the parts are used for. They've given us sketches of the parts [product information], but you look at it and say what's that? The company has mentioned it in the past but never followed through on it – that's when they give us these stupid sketches. It would be good for everybody to go and see. (Jodi)

Overall, workers suggested that opportunities for training within the production process are sometimes lacking. While there is limited opportunity for upgrading skills and knowledge – workers doing so remain frustrated by the structured hierarchy of work that does not allow them to use new skills, knowledge or insight. Worse still is that particular groups of workers have less access to courses in the first place.

Work-Based Informal Learning

Probably the most highly regarded learning activities that workers reported were those they did outside of courses and formal schooling. This learning was done alone or with others, but its most satisfying and effective aspect seemed to be that it was controlled by the learners' individual and collective interests, styles, schedules, perspectives and abilities. One worker captured the collective and self-driven nature of this learning with these words: "Learning is ongoing, driven by interest and curiosity ... you learn by asking different people at work or in the neighbourhood" (Daniel).

In the workplace, those interviewed revealed a wide range of work experience that is international in nature. A few people have worked at the same type of job for a very long time, either within the workplace or others like it. They also talked about gaining abilities, knowledge, experience and skill in jobs like building demolition, auto painting, sales, cook, nanny, heavy equipment operation, waiter/waitress, upholsterer and more. As many people on the floor recognize, there are diverse pools of life experience, skills and knowledges within the local membership.

Workers discussed the importance of learning outside of the formal school context. Informal learning opportunities were seen as the most valuable types. The two workers below (male maintenance and set-up workers respectively) outline the importance of these learning opportunities:

> I started a bit at a time, I learned about the all sorts of things. But by trying, I improved. Not because I went to school though. See, the beauty of this place is that I had things I could learn, so really, and I never needed to go to school to learn ... because I had the opportunity to go on and learn more here ... it has been the best period of my life. (Lew)

> I had the opportunity to try things. You learn, not from books, you learn from life. You do that, you make a mistake, next time you try to avoid the mistake, sometimes you make a different one. You try to improve yourself until you master it. It stays in your hands, your system [but] you have to use it to keep it. (Eduardo)

Despite the barriers of communication with a multilingual workforce, there is still a great deal of informal learning. Workers in all areas of the plant, even those who appear to be most restricted in their access to new skills, search out and find the time and space in which to talk about, and experiment with their work. In Mary's experience,

> [T]hey leave you alone after a while and you try to learn by yourself ... Some places they have qualifications, but not where I work ... They don't care about what kinds of qualifications people have if they need people, as long as the people learn.

Often the production operators' central learning experience at work involved health and safety, keeping the body from injury, and finding ways to "relax" a bit.

> Basically everyone has their own ways of working in there ... "Let's see if we can do this faster, save some time" and so maybe that way we can sit here and relax the body a bit. (Rava)

> One day a machine caught on fire. I know how to use a fire extinguisher because my husband showed me at home. I shut off the fire. They have a fire extinguisher but nobody showed us how these things work. They ask us what we want to learn, what we want. But if you don't give an idea they don't know nothing about that. (Mary)

Important job training like health and safety, is often done by and for the workers themselves, informally. This acts to fill in gaps where the company's initiatives fail. The above quote from Mary highlights another important point as well. How are workers supposed to know what training they need if they cannot get a clear idea of the production process as a whole? The company puts the responsibility on workers to indicate what they need to learn, but this may not be the best approach without general training about production processes. Several workers made suggestions that point out the need for a comprehensive understanding of the production process, something that is limited to those with more authority to move around in the workplace.

> In order to help the only thing I can suggest is to have good courses in the workplace to explain what people are doing in the workplace and then people can get interested in the jobs and maybe something can come out of it. (Lew)

Throughout the interviews, workers gave examples of the many different ways they learn best. Some prefer to read and do, some talk with others, watch then do, or some combination of these. A dominant theme is a preference for "active doing".

> *Interviewer*: What's your best way of learning?
> *Jodi*: Depends on what it is. Probably the best is having something to read and doing at same time, so you can check back and read about it. Ninety percent of what I know about the job I learned at work *[but]* I have manuals at work for procedures too.

Although some workers still prefer to do their learning alone – in most cases learning is done best collectively amongst other workers. Learning with co-workers may be preferred because they share a similar perspective of the work process and often have similar interests at heart.

The Cultural Historical Dimensions of Learning

The CHAT perspective draws our attention to several specific features of work and learning at this small parts manufacturing plant, some of which have been introduced already and others that will be discussed in the following sub-sections. CHAT's key contribution is to direct our attention to how these seemingly extrinsic features of learning are actually definitive of it. Learning is deeply affected by the position of this firm in the auto parts sub-sector as a small and vulnerable Tier 2 supplier. Cycles of lay-off and recall shape not only what kind of learning can take place at work, but also the characteristics of the workers that are employed. As in other sites, the specific organization of the labour process is crucial. Continuous

run assembly lines, in particular, narrow the ability of workers to participate with each other and hence shape the learning at work in a narrow way.
In contrast, the labour processes of trades/maintenance, as well as set-up
and even material-handlers, have openings for discretionary movement
and autonomy and hence contribute to richer learning experiences for
these workers.

Language is a key mediating tool or artifact that shapes possibilities for
learning and how learning is accomplished. But more than this, the mediation of activity by language is a contested site of struggle between workers and management. Formal and informal organizational policy relating
to language thus shapes many possibilities for workers. In addition, as in
some other research sites, the practice and developmental trajectories are
made more complex in this work site by multiple social standpoints
mediated by key work-specific rules and divisions related to occupational definition.

The Divisions of Learning at Work

Many literatures have established that the organization of work, including
its formal rules, divisions of labour, standard operating procedures, job
descriptions, managerial rights and practices, labour relations regimes,
technological change, normative structures, different cultures, etc, all play
important roles in defining what goes on in the workplace. However, our
understanding of the relationship between the organization of the
workplace and workers' learning specifically is much less developed.

We see in the descriptions above a glimpse of the social organization
of the workplace where movement, the discretionary control over time and
participation are structured via the placement in the organizational structure. However, and this is the perspective typically missing from most previous research, we see a production worker's view of this process and its
relationship to learning from which we can begin to understand the types
of agency, limits and pressures that their lives actually entail. In general,
there are clear limits to worker involvement in work processes shaped by
the organization of work. For production workers at this parts factory, for
example, intensification of work and the demands of the assembly line
diminish human energy, limit time to reflect, constrain access to an experience of the broader production process and restrict discretionary interaction among co-workers. Through the organization of work, the production
worker comes to participate in work-based learning life more as an "operative" or "tool" than as an "agent" of production. The interviewees supply a description of the link between context as well as learning and cultural and material dimensions of power and knowledge. By looking at

work and learning in this way we take an important step from the local production of experience to (aggregated) macro descriptions of work and society.

Power comes in many forms. In our view it is rooted in, but is not limited to, the distribution of material resources. It also has cultural and symbolic dimensions, it is mediated by cultural and symbolic tools. Many of the earlier quotations speak about the ways that workers' experiences are shaped, distributed and controlled. Some interviewees have spoken about the symbolic power of credentials and a countervailing valuation of credentials that privileges informal learning and the achievements open to workers (raising a family, learning from co-workers despite resistance from the organization of work, etc). Credentials were understood as an important element in the struggle for legitimation and control. Specifically, workers describe that the organization of the workplace is intimately linked with the ongoing maintenance of managerial control over knowledge and power, and that these practices also included the maintenance and struggle over symbolic forms of capital as well as material resources. In the preceding profile section it became clear that informal, everyday learning was central to how work was done, but it was also the means by which workers could, in some cases, resist the formal rules and imposed organization of work. In the case of the latter, discretion and autonomy were key. Thus, probably the most important point introduced here regards the degree to which different workers have autonomy within the workplace, certain levels of discretionary control over their time and space, and hence varied opportunities for discretionary learning in the workplace.

In terms of this work site, when workers have more decision-making power over their work they seem to become more skilled, and their overall satisfaction at work increases. When workers do not have this significant freedom they still find small gaps in the organization of work to consult, reflect and share experience among co-workers. But winning these small freedoms is a struggle. Indeed, as the freedoms become smaller, the struggles become directed not towards greater skill and knowledgeability but towards achieving basic physical necessities such as being able to go to the bathroom when needed (Gina) and simply being able to rest one's body (Rava).

Struggles for small freedoms made up much of the learning and work lives of production workers in the factory and, to a lesser extent, the material-handlers and set-up people. These people, with varied success, had to seek out and actively create opportunities to learn from each other. However, the case is somewhat different for the small number of quality inspectors and the trades and maintenance workers. In the following comments, note the importance of "free time" for workplace learning during a talk over coffee in the morning:

> Learning about equipment like the "Shadowgraph" [quality control equipment] we do on our own ... We sit and talk in mornings at work. This is valuable learning time. (Anne)

Trades and maintenance work, on the other hand, is designed to include a great deal of problem-solving and learning on-the-spot in terms of production so certain forms of informal learning are structured into the job itself:

> *Interviewer*: Did you learn that through training, from friends?
> *Lew*: Just stole it during life. Not so good the first time, second time you're OK.
> *Interviewer*: Did you get a chance to try out what other people were doing?
> *Lew*: I started in general maintenance, and was able to look at things. I worked in machine shop, learned a little bit at a time and learned to build parts. I improved over time. Not because I went to school. Only went to school a couple of years in Canada to learn English. Learned some from other people, helping, learning, then doing by myself. No kind of program. At that time, I worked with a really nice person, a European. He taught himself a lot of things. I don't even know what kind of qualifications he has.

Because of the customized nature of some of the tool-and-die work, the trades workers at the parts factory have relatively more learning opportunities related directly to their jobs, than do other workers in the plant. In fact, historically this type of autonomy has been as much a conscious struggle to achieve as it is a causal effect of the nature of the work. But for our purposes, the organization of these people's work leads to distinct perspectives on learning that were far more positive. This positive sense of learning, and in several cases technical schooling, was related to the ability to actually use the formal schooling in some way. "People need to put together theory and experience in the workplace. Theory is okay, but it doesn't stick with you unless you try to put it in practice ... then you can attach the theory to the job you're doing" (Lew). The maintenance, trades and set-up workers are able to learn more because they can, "see the whole process," have more control over their time and more opportunity to practice what they learn. The production operators did not have the same opportunities.

> *Rosa*: Before they used to take us on a tour. When you work all the time you don't see that much. They don't do that anymore. You learn more

and you see more, when you watch where you work, you don't think so much because you have to think about what you do. When you're working you don't get time to think about your workplace. They said they'd do it before but then they stopped.

Interviewer: Do you get to see the whole process?

Rosa: Not really, I don't see everything. The parts through, the plastic, pieces of plastic they put in the machine. They melt. The company has lots of experienced people, engineer. They have lots of training. I work on one thing all the time.

While the opportunity to learn new things can be limited within the workplace because of the way the work itself is structured, there are also barriers at a broader level, and this is where learning opportunity, according to those interviewed, plays such an important role.

Promotional Practices

Issues of learning opportunity are critical to understanding what workers told us in our interviews. Building on this discussion of autonomy and discretion, we saw that basic relations of workplace power and the control over workers' experience and learning were closely related to promotional practices. In the parts factory, promotions (paradoxically) were the means by which one obtained the skills apparently necessary for promotion. Promotion expanded one's informal learning opportunities through an expansion of discretionary control over time, space and social participation. As the central means by which managers regulate workers' experiences, the promotional process (the sole discretion of managers) contributes to the production of a segmentation of learning in the plant and in this way contributes to reproducing a range of class, ethnic and gender divisions. Using credential inflation provided the opportunity for managers to generate a veneer of objectivity that served to legitimate control over workers' experiences, but this control was important to a more general control of workers *vis-à-vis* power/knowledge relations. Thus we have a relationship between segmented internal labour markets and learning. That is, learning is one of the means by which the internal labour market is segmented along the lines of familiar social divisions. Gender and race are both implicated in these processes. In terms of gender, the division of informal and formal pathways to promotion is integral to the overall gendered division of working life on the shop floor. It is reproduced continually through the interplay of opportunity, learning and the formal and informal rules of the firm.

For example, while one of the female (Canadian-born, white) production operators undertook community college credentials to gain promotion to a quality inspector position, a male material handler watched, chatted, and quietly learned the set-up job. Similarly, a male set-up worker described how he spends time asking questions and observing the work of the tool-and-die repair worker (also male) whose job he would perhaps like to do one day. Based on this, and often a supportive report from the senior worker, the men are promoted to the set-up and tool-and-die repair "trainee" positions respectively. These are informal routes. They are also male-to-male learning relationships, which tend to reproduce a gendered division of labour in the workplace. Ricardo, a male set-up workers, recalls:

> I started from the bottom. I swept the floors and then I saw a chance to help on the line. I learned about it. Then there was a posting there for set-up training. They give me a chance. They teach me more there and I learn the job.

Another (male) material handler tells us how he is planning to gain promotion: "On breaks, at the end of work, [the set-up people] teach me how to set-up machines, change die, clean the machine" (Marco).

There is evidence of systemic gender discrimination in some women's experience of job mobility.[6] Contrast the "natural" job movement above with the experience of two female ex-production workers who have become inspectors:

> When the men go through they say, well, he's done this or that, or he's got a certificate, but a lot of the men they hire didn't even have the courses they required us [women] to take. (Anne)

> I've had the opportunity [to get a promotion in the workplace] but just before I got my job in the department there was someone hired off the street. We used to have [different types of] inspectors. They wiped out [some types and now] we have the backup quality inspector who is actually paid the operator wage ... And the [guy] in one position now, they posted for one level but hired him at a higher level, and I find you really have to bug them to get promoted from within because a lot of times they'll put a posting up that nobody can [apply] for anyway because they don't have the qualifications for the job. It's almost like they post like that on purpose so they don't have to train anyone from within ... You end up having to get the training on your own. (Jodi)

Thus, the ability to use social networks for learning about a new job is linked to the amount of time and freedom of movement a worker has around the plant in the first place, and gender. In this respect, the differences between production operators on the one hand and material-handlers and set-up workers on the other is quite significant.

This division in workplace labour reflects much larger and deep-rooted societal issues of training and division of labour, but several workers commented on this division in their workplace specifically as being part of a workplace culture.

Jodi: I find our company is very male motivated. Their attitude, they come across like women are lower and dumber, and if you're a male, you're a great guy and you can do anything and they promote them.
Interviewer: Do you think that being smart or dumb comes into that?
Jodi: No but that's their attitude. We have one girl who worked in shipping/receiving, really had to fight to get that job. Guys hassled her about everything if she wasn't on top of things. Guys could talk for half an hour and didn't get hassled. No woman material-handlers, only two guys working on the punch presses, the rest are all women. No women in set up or tools. In the quality area, there were more women than men, but now with people going off or whatever, they've tried to hire more men. This woman *[from shipping/ receiving]* would put grievances in, and she'd win them, but this didn't change the daily environment. All it does is that the supervisor might give her a written apology.

Access to training and promotion hits hardest those workers with the least formal credentials. In the case of this plant, this means women on the production line.

Discourses of Learning

Many of our interviewees expressed much satisfaction with their learning lives generally, particularly older workers who stressed their accomplishments surrounding establishing a home and family (e.g., Rosa), and trade workers who consider their training to be constant and highly satisfying (e.g., Daniel). However, others (mostly women and production operators) would like to be learning more in the workplace if the opportunities were made available. Again, those who appear to have been most frustrated in their learning efforts were more likely to be women who have attempted to move off the production line. "I always felt frustrated that I couldn't learn more, I always wanted to learn, always, always, always" (Anne); and "Employers want to keep workers exactly where they are. Don't want them to better themselves or learn more unless it's one or two chosen individuals" (Jodi). Related to this is the notion of a specific way of talking about, engaging with, and acting upon the world in the context of a certain conception of "learning." Workers' engagement with this discourse helps shape the options that they feel are available to them. Moreover, this discourse and the positioning of oneself within it is directly related to one's occupational position.

Informal types of learning, especially the "hands-on" and collective aspects, are not what we are "taught" in schools to call "learning." Ask someone about learning and generally the answer will focus only on the courses that have been completed, not the hours that have gone into learning to be job experts, learning to be parents, learning home-plumbing, learning how to deal with the local town council, etc. There is a bias in the way we have been taught to use language that helps keep this learning hidden and this makes it difficult to talk about. Learning among workers in an industrial organization like this compounds the problem. There is little recognition of skills mastered and expertise gained. Promotions form one of the only legitimated forms of acknowledgement. While informal learning of all kinds is surely not limited to workers alone, due to the relations of production and learning in the workplace it is workers who are left with the fewest tangible forms of acknowledgement. An effect of all this is that workers will generally underestimate the amount of learning they actually do.

However, an important exception to this effect of the "language of learning" can be found among the trades and maintenance workers who seem to have less of a problem identifying their learning *as* learning. One main reason for this seems to be their experience with their trade apprenticeship. In this type of education (as opposed to standard, classroom-based education) hands-on experience is fully credited and seen as the core of trades certification. This gives these workers some alternative experiences of their own from which to make different meaning out of the terms "education" and "learning." While the work of the trades and maintenance workers seems to involve learning on a daily basis, the time spent learning amongst other workers is more hidden. One maintenance worker said that for him learning "is a part of every single day" (Frank). Likewise, quality inspectors must master a great deal of equipment on an ongoing basis as well as understanding the larger production process. Workers like this material-handler, despite his description of many different learning activities involving set-up, fork-lift skill/practice and cleaning, claims he does little learning: "It took me two days to learn my job" (Marco). Production operators said that their job was routine and offered little chance for discretionary learning.

What do workers mean when they talk about workplace learning? The answers suggested by research at the parts factory are that they mean the things that shape participation including the specific rules, policies and practices that limit and produce openings for discretionary involvement. But despite the challenges of learning at work, openings for resistance to limited involvement are detectable. These can be understood with reference to "strategic" versus "tactical" elements of power and practice (DeCerteau, 1984; Sawchuk, 2003a). In this regard, the former represents practices beginning from a position of relative power; they depend upon some level of control over time and space. The latter, tactical practices, operate under conditions of discipline and control. As our data indicated, tactical activities included workers gaining experience about the workplace talking to other workers on break time and on one's own non-work time; informally practising the non-English languages of others; applying practice from outside work in the workplace (e.g., fire extinguisher use, first aid), etc. For the female operators, however, even the most modest of tactical achievements in the workplace (e.g., discretionary bathroom breaks) were limited via the structures outlined.

The male tool-and-die workers exemplify a type of practice that is tactical, but with significant strategic elements that may exemplify the importance of seemingly insignificant features. In general, trades workers in Canada occupy a favourable position in the labour market, but they are also seen as essential for the production process and difficult to replace. During our research, despite frequent lay-offs of other workers, no trades workers were ever laid off. Thus, (all male) trades workers actively develop their relatively free run of the entire production process; physically "private" social space (the tool shop); a specialized technical language; learning histories that anchored alternatives to the dominant/dominating discourse of (formal) learning (i.e., apprenticeship); and extra free time from housework before and after work (and possibly in the production of their credentials as young men). These tactical/strategic practices of control are experienced as moments of enhanced collectivity (among male workers) though this is problematically narrow (thus partially ratifying other sets of discriminatory power relations).

Discretionary Time for Learning and the Home

Many of the themes in this section are contained in Chapter 8 but it is important to note here that, while gendered relations are visible in this paid workplace, differential access to promotional structures, and activity outside the workplace has an important effect on learning as well. Specifically, gendered distribution of free time after work shapes training opportunities which in turn affect opportunity in other spheres of activity. According to these manufacturing workers, the female "double day" (and the associated lightened non-discretionary workload for male partners) has a devastating effect on time and hence participation and learning. Indeed, in setting up educational programs with the union over the course of this research, it became clear that the women workers had difficulty attending regularly; often having to rush home to prepare meals, clean house, care for children, and in some cases even go to additional part-time jobs. In general terms, these types of class/gender effects (professional and upper-class people hire help or can afford to have a single income) shape learning and can only be overcome with extensive planning and great investments of energy.

> I have a really cramped time, so many things I love to do, but I have the children, I get up at 5:00, don't get home until 5:00, so by the time supper's cooked, and I've played with kids, and get 8 hours sleep – I'm really squeezed for time. I have to read on the bus ... [for female workers at the plant] to continue their education in the evening is such an inconvenience for everyone: they're suffering with guilt, the children are sick, he's not a great caregiver. I'm wondering if he's looking after the little one with a fever. I'm studying for exams, no cooperation from my partner ... For a lot of women, they need that support from a partner. If they have workplace training – especially workplace training – if workplaces would just squeeze some time in and allow the employees to learn, or take a day off and allow them to learn, or have more options for Saturday trainings, it would be a great help, especially for women. (Jodi)

Learning, particularly for working-class women, is marked by contingency and the need to create space for learning within narrow openings in the day/night. "I have spent my whole life working and saving money! When I'm not working, I'm cleaning the house, or cooking, or shopping, or going to the bank. I have no extra time!" (Rosa). The gendered division of labour affects workers and their learning in at least two major ways. First, female workers have less time for free association and casual conversation than male workers because of their job. Production operators (mostly women) are closely tied to a continuous production line unlike set-up, tool-and-die makers, etc. (mostly men). Second, a gendered division of labour on the home front demands a "double day" from female workers with families, through an unequal splitting of childcare and housework. More than one female worker (all with employed male partners) spoke of homework cutting into both time and energy, in some cases affecting attempts to take courses. Ideally more formalized learning should take place on work-time thus helping to relieve the burden of unequal divisions of labour at home. However, even training done at the workplace before or after shift, as some workers pointed out, would at least eliminate valuable travel time and the need for alternative transportation arrangements.

The male workers interviewed were not unsympathetic to the plight of their union sisters. In many cases they recognized that the division of labour at home and work was a problem and things should be changed. However, few had any constructive ideas on solving the problem. Other men commented on the division of labour in patriarchal terms inspired by an idea of masculinity that deflects and severely limited any solution of gendered inequalities.

> Some people are happy with 10 bucks an hour. For girls it's a different story and for guys it's different. For simple things no one's going to pay you more than 10–15 bucks. For me, I never was a chicken, if you are a man – you have to work a little bit and make more money. Some people are lazy. (Vlad)

In working closely with the union local over the course of our research we suggested that overcoming the division of labour for female workers, particularly with regard to job training, would partly depend on the ability of the union to generate a brand of solidarity that deals directly and aggressively with patriarchal views of gender relations. Union-led courses would be ideal for these purposes.

Relations of Language in the Workplace

One of the more complicated structures of control that mediates participation and learning in this workplace involves ethnicity as it relates to (written and spoken) English language fluency. Again, these processes shed light on how internally segmented labour markets are achieved. As a linguistically diverse workforce with first languages including Italian, Portuguese, Croatian, Polish and Spanish as well as English and French, language could be expected to play a significant role in mediating activity at work. Indeed, the role it plays is not all that positive. We should recognize, however, that these apparent difficulties are actively produced *vis-à-vis* the organization as well as the formal and informal policies of the workplace. Difficulties related to ethnic division between workers, disempowerment because of lack of fluency in English, limited opportunity related to language learning, and so on, are not a priori, natural or given. At the same time, however, from the workers' standpoint, the notion of working with others who speak a different language was viewed less as a problem than an "opportunity" to learn a working knowledge of another language. Virtually every interviewee who spoke about working closely with someone who spoke a different language did so in a positive way. Indeed, most, saw it as a way of combatting the monotony and routine of the work day. People were proud of their working knowledge of different languages. The shared experiences of production problems served as an important bridge between linguistic divides constituting a shared object of concern, ready-made subject matter for language learning. Of those we interviewed, many had developed functional literacy in two, and sometimes three, other languages, despite the fact that to do so sometimes required them to break company rules by leaving their work station to have a conversation with a co-worker who spoke a different language.

Opposing this rich learning opportunity, the company attempted to impose English as the sole language in the workplace, using it for memos, posted notices, supervisory information, newsletters and equipment manuals. Language diversity as a "resource"[7] was transformed by management and the policies of the workplace into language diversity as a "problem" to be solved. In other words, the need to satisfy the requirements of (English-speaking) management turned the otherwise positive potential of this linguistic diversity into a significant barrier for workers. Combined with the other patterns of division based in informal interaction, language relations act as a significant structure of workplace experience by controlling access to textual information at work, formalized training, promotion, etc.

> *Interviewer*: Am I right to say that people in the plants are not likely to be attracted to courses?
> *Jodi*: Right, there are a lot of language problems. The number one thing that people will say is "I can't speak English very well," or "I've been out of school too many years." But a lot of them are quite capable of learning given the opportunity and a little encouragement. *[But another barrier is that]* if they have the desire, time for travelling is a problem, so it would be better in the workplace.

Despite the resources and willingness for a different (multilinguistic) flow of information, informal training and problem-solving, the institutionalization of English as the privileged language of work activity produced differentiated and differentiating forms of participation and learning among workers.

English is the language that supervisors, management and office people use; it is the language of all written communication from the company; and is used in training courses. Where workers speak other languages, management/supervisors "reorganize" to turn ESL workers towards English.

Although informal, experiential learning is a valuable contribution to English-language skills, further support would help ESL people, workers say. Extra English-language skills would allow workers at all levels to more ably represent themselves in "official" interactions (with management, supervisors and co-workers). Multilingual translations of workplace communications (verbal and written) would be helpful but workers with English as a second language are committed to learning more in the future.

> My mother held me back from kindergarten so that I would be able to learn English good enough. (Marco)

> Most of all I would like to speak better English … If they offered courses people would really be interested. I want it for the job and for myself. (Ricardo)

Some workers are actually going beyond the English-only learning to understand the language of co-workers better. Mary's first language is Italian. "I watch Spanish soap opera. I talk a lot with girl at work who is Spanish. You have to go slowly. Talk about the soap opera. The Italian women joke with me, first English, now Spanish! I'd like to learn more if I could. Also, Portuguese, I can understand if they speak slow" (Mary). Family members are usually important in aiding the learning process.

> *Interviewer*: What do you do when you have to do written stuff?
> *Lew*: When it's really necessary I ask my son and my daughter to write what I want to say. But I don't really have much communication to exchange. In my second job, I was a foreman and kind of needed to write. I kind of learned what I needed. You know even if I needed something for my line. I had to make a requisition. I learned to write whatever I needed for that. But then when I left there, I didn't need it anymore, so it's gone *[laughs]*.

Support for learning English was thought by almost all interviewees to be an important element for full participation in workplace life. While we situate this discussion of recommendations in the last chapter of the book, it is clear that this support could take on a number of different forms that go beyond teaching the individual to learn English. For example, some workplaces have produced written information in the language(s) of their workforce. Peer support could also be implemented to facilitate writing for workplace documents. In the absence of appropriate support, it is clear that workers will have a more difficult time communicating among themselves and with management.

This issue became particularly visible in discussions with trades workers about the "problem" of different languages and the imposition and power of managerial English. In these discussions, it became quite clear that knowledge and power mediated relations of language. Take, for example, the comments made by this (male) tool-and-die worker. Similar to most of the (predominantly female) production operators, this tool-and-die worker has limited abilities in spoken, and especially written, English. For production operators this limitation is a barrier to promotion and a source of managerial control, but the relations of language play out in a significantly different way for the tool and die worker. When asked in his interview about language as a possible barrier in the workplace learning, he responded, "The machine shop English is specialized language. I don't have any problems dealing with supervisor [because] ... the supervisor doesn't even know about our tools and I have to explain" (Vlad). When he discusses the deficit that language can produce for workers (in relation to management), neither English nor Vlad's native Croatian language is referenced. Instead he discusses the specialized technological language of tool-and-die work. Thus the notion of the language barrier is shown to be intimately tied with positioning in the everyday flow of information, specialized knowledge as well as the resources and authority to develop it on an ongoing basis in the workplace. Trades worker such as Vlad are simply not tied to the organizational texts and the need to speak English in the same way as the assembly-line workers. This positionality, based on information, knowledge and learning, acts to shift the balance of power and control and produce radically different results for objectively similar language contexts.

Concluding Remarks

This chapter has reviewed data from a work site in which learning is both a source of satisfaction for workers and a site of struggle based on a variety of interlocking factors. Light manufacturing, and specifically the small parts sub-sector of the Canadian economy, provides a complex backdrop for work and learning. Moreover, the specifics of the firm, the managerial strategies, choice of technologies and labour processes as well as formal and informal organizational rules come together in unique ways to give the learning of these workers its shape and pattern.

This context is considered adaptive rather than innovative. It narrows the opportunities for expansive skill and knowledge development because it must rely on outdated technologies and outdated control structures stemming from traditional industrial labour processes. At the same time, these matters are not experienced uniformly by different workers. A central theme of the chapter is the concept of divisions: those between worker and management and those between occupations, genders and ethnic/linguistic groups.

This chapter has added to our analysis of class divisions in work and learning, and how these are accomplished in the workplace. However, it has also made a distinctive contribution to our understanding of how learning and work are differentiated for workers based on gender and ethnicity/English fluency as well. This differentiation occurs through conventional forms of exclusion, for example, via differential access to courses. However it also happens through the various informal pathways of learning opportunities that are interwoven with mechanisms of job design, promotion and even mediated by, among other things, the very way people talk and think about learning.

Notes

1. Personal communication with Human Resources (October 1997).

2. The report also claims that there is parity between management/professional employees and production employees, however data consists of "available programs" rather than participation, length, frequency, etc.

3. Tier 1 supplier refers to companies that supply complete component systems while Tier 2 suppliers are those that supply parts of these component systems, sometimes to auto assembly companies and sometimes to Tier 1 suppliers.

4. The ongoing program of using workers' expertise produces results such as the ones documented in an issue of the company newsletter during our research. In this, a worker's idea was indicated to have added valve to the extrusion head operations saving the company $13,000 annually. The worker was awarded a one-time cheque for $284.40 (about 0.7 percent of the savings he generated).

5. Near the end of our research at the site, according to union representatives, the company commissioned a management consultant who delivered "culture" and "strategy" courses to management. These courses are presently aimed at moving the company toward a more intensive "Learning Organization" model. This approach demands an ever greater integration of workers' knowledge and expertise into the production process which, as noted above, helps explain the company's interest in educational programs.

6. Systemic gender discrimination is also reflected in the generally lower wage levels of both (female dominated) inspector jobs and production line workers. The inspector jobs appear to require more formal credentials than male dominated "set-up" jobs yet are in a lower wage category.

7. For a related discussion of these issues see Ruiz (1984), Willinsky (1998) discusses the imposition and use of English in the broader context of Canadian "nation building" and assimilation.

CHAPTER 7

Garment Workers:
Learning Under Disruption

co-authored with Clara Morgan

Pramila is a packer in a distribution centre for a garment company whose workforce has been drastically reduced from over 200 to less than twenty workers. The company no longer does any production on site but relies on small contractors. It employs only pressers, general labourers and packers in a small, crowded building without air conditioning. Pramila is in her late 20s and has some college education. She emigrated from India in the late 1980s with other members of her family, including her father who is active in the steelworkers' union and her husband who also works in a garment factory. When she started working for this company shortly after arriving in Canada, many Indian women were working there. They soon chose her as spokesperson and shop steward because of her leadership skills, developed in high school, and her relatively good English. She speaks passionately about language barriers for immigrant workers and apologizes for her own limited English. But she expresses herself quickly and clearly and stresses how union courses have improved her English.

Pramila has taken ESL classes and a basic computer course as well as several union education courses. She constantly refers to the difficulties of immigrant women workers' overcoming language barriers, poor wages, housework duties and unsupportive employers and husbands to attend courses, as well as her own more recent difficulties in balancing childcare and making a living.

> *More and more people are being forced into home-working with very poor wages and no benefits. The way to go is to get work out of homes, out from" under the table," and into factories. Most home-working is done by immigrants whose main concern is to make money and for the future of their children. They don't have time for courses or for much else.*

At the factory, Pramila received informal job training on sewing machines for a few months but she was not a fast sewer and they made her a packer. She learned to do union work mainly through contact with local leaders and learned about current affairs and how to adjust to family life in North America through watching television and community observation. She is now reading everything she has time for and consulting family members about early child care. More generally, she talks about how she has constantly learned from other people's experience and how this informal learning has led to changes in her own life and actions to help others.

> *When you learn, you meet different people, they tell you their own experiences, what's happening. So, then I check with myself if it's happening to me. If some people have better working conditions than me, then I think this should be changed for myself and for other workers. That's the way I learn. I feel.... Like a grievance case for recent immigrant workers who went on vacation and the company didn't want to take them back. These kinds of things happen all the time in our workplace. That's how I learned.*

While Pramila's story helps introduce important dimensions of the garment research, the male worker' experience is often quite different. David is a middle-aged cutter. He learned the trade from his father who was a tailor in North Africa before the family immigrated to Canada in the 1960s. After working as a sewing-machine operator for a year and going to night school to learn English, he has worked most of the time as a cutter. He continually talks about how hard it is to find any job at all in the industry. He now works in a small women's clothing factory which employs mostly female sewing-machine operators and a few other male cutters. New computerized machinery will mean less work and less pay. He took a computer maintenance training program while he was laid-off for a few months a year ago but returned to the factory immediately when he was called back. He originally became involved in the union when the first company he had worked for owed him severance money and he got to know union activists. He later served as a shop steward but quit because he became resigned to the view that workers can not really do much to improve the workplace.

David did not like school but he now enjoys studying. It keeps him busy and diversifies his skills. He doesn't think his chances of getting a job in computer programming are great but believes he has to learn it to have any chance at all of a job in case he loses this one. He was prevented from taking a union workshop by his boss. He thinks workers should not wait for employers to offer training but should go out on their own to develop their skills, but that unions should develop classes for workers to talk about new skills and techniques in the industry. David has informally learned a variety of cutting techniques for different forms of garments. Beyond his job, he spends a lot of time watching educational television and reading about religions, insects and animals. He feels his job is unchallenging but relatively secure.

> *I'm using 10 percent of what I know. But I am making a salary. Even if I were offered a job using more of my skills, I'd be afraid of giving up this job to take it.*

Introduction

The garment industry experienced a massive upheaval between the late 1980s and the mid-1990s. Over 30 percent of the jobs in the industry were lost. In Ontario almost 40 percent of jobs were lost, while in Toronto, the original production centre, about half vanished (ILGWU, 1995). Over a quarter of the firms disappeared by 1992 and smaller firms and contractors relying on unorganized home-workers have since proliferated (ACTWU, 1995). At the same time, imports increased from under 30 percent to over 40 percent of the apparent Canadian garment market. Global garment production has been shifting to lower-wage developing countries for decades, but garment imports from the United States increased sevenfold between 1988 and 1992 under the impact of the Free Trade Agreement, which accelerated movement of large U.S. garment retailers into Canada (ACTWU, 1995). As one of our key informants in this sector explains:

> Imports are our greatest problem. Imports are going to drive the industry even lower to a minimum wage labour force. In the past, we had sewers making $15/hour, cutters making $22/hour. Those jobs will be gone. In negotiations you hear all the buzz words like 'we have to be competitive.' They're asking for concessions and rollbacks. The ones that are surviving are accessing the U.S. market. If they have to rely on Canada, they won't stay in business.... The plants that we are going to keep are the ones that will be automated, where people are cross-trained, they are highly flexible, highly productive.

Computer-assisted design, marking and cutting equipment, automated spreading machinery and programmable sewing machines increased production speed and reduced employment but the industry has remained highly labour intensive. Sewing still accounts for most labour costs, garment manufacturing is still largely based on one machine/one operator, and the soft, pliable nature of cloth continues to limit automation (ACTWU, 1995).

The industry has become highly fragmented, with a few large integrated manufacturers and many small ones relying on adaptive low-wage strategies of contracting out most of their production (Lazonick, 1991). Men's suits are often more standardized and therefore easier to make in the plant. Men's suit manufacturers increasingly invest in capital-intensive technologies (e.g., CAD/CAM and overhead conveyer belts) but maintain strict control over the work process by applying "Taylorist" methods (ACTWU, 1995). However, most other manufacturers increasingly opt to centralize their cutting operations and contract out their sewing operations to very small factories (ILGWU, 1993, 1995). Many manufacturers have left the heart of the Toronto garment district located at the Spadina-King juncture and spread into outlying areas including Scarborough, North York, Etobicoke, York and East York. Another key informant summarizes the new situation for the majority of small employers:

> Garment production is not done in one factory anymore. Cutting is in one place. Contractors are sewing in little factories all over the city, thirty people here and there. Contractors don't have big equipment. They can shut down and move overnight. Small contractors have only sewing machines which they can load up in the back of a truck. Companies will break out operations in different plants to split the liability of the business. Some declare bankruptcies, avoid paying wages and benefits, and then open up somewhere else. They take some of their people with them and they haven't lost a day of operation.

Contracting out to home-workers[1] is difficult to measure but is estimated to have grown rapidly (Dagg, 1994). Home-working has existed for over a century as a flexible form of employment in the garment industry. As Steedman explains, the condition of home-workers during the 1890s "placed them in a weak bargaining position and allowed the contractor ... to maintain a downward pressure on their wages" (1986: 156). The same pressures exist today for women who sew garments from their homes but are not justly compensated for their skilled work. A resurgence in home-working in North America has been documented since the 1960s as garment manufacturers tried to compete with the effects of trade liberalization policies and cheaper imports from Third World countries (ILO, 1995). Home-workers are generally among the most vulnerable workers, without access to either basic employment standards or collective bargaining. A 1991 study found that Toronto garment home-workers made less than minimum wage with no benefits, paid their own equipment and operating expenses, worked very long hours and suffered serious health and safety effects from their work (ILGWU, 1995). In 1983, garment workers in Montreal revolted against the rise of home-working and contracting shops with sweatshop working conditions, and the against passivity of their union by "unleash[ing] the first industry-wide strike in forty-three years" (Lipsig-Mummé, 1987: 41). But workers' collective action has become increasingly difficult as the fragmentation process has proceeded. During the early 1990s, the unionization rate fell at twice the rate of industry closures (Dagg, 1994: 9).

The industry has become increasingly like a hollow pyramid with just-in-time production by workers in small factories and home-workers as the key component.

> The hollow pyramid is organized with five major tiers. At the top are the retailers and large super-label manufacturers. This tier is heavily concentrated or centralized. The top tier controls production chains of either the retailers' own labels (a private label) or a well-known brand or licence (super-label). The top tier controls designs and marketing as well as distribution. The second tier consists of jobbers. Once manufacturers, now they mainly organize production. Contractors who cut fabric or do some sewing make up the third tier. The fourth tier consists of subcontractors who sew the garments. The fifth and final tier is made up of home-workers – former factory workers who sew in their homes." (ILGWU, 1995: 35)

In reaction to retailers' demands for quicker response times, manufacturers have purchased new computerized design, marking, grading and cutting equipment for the production process. Sewing operations are separated out and moved down the pyramid to contractors, subcontractors and home-workers. Design and cutting are linked to retailers with advanced information technology while sewing is completed by low-wage workers with poor working conditions. The persistence of labour-intensive production, combined with the proliferation of smaller factories and subcontractors in a highly competitive environment, has encouraged reliance on these adaptive low-wage managerial strategies as well as provoking high turnover rates. In fact, few previously unionized factories have implemented innovative capital-intensive strategies such as modular manufacturing and most have followed some combination of the following: factory closure, downsizing production lines and relying on imports, decentralized reliance on contractors, driving down labour costs, reducing in-house design capabilities and little investment in new sewing technologies (ILGWU, 1995: 40–41). This is a classic case of the adaptive managerial strategy referred to in Chapter 1 by which organizations attempt to survive predominantly through increased productivity based on greater coercive pressure on workers.

Since the mid-1990s, official employment rates in the garment industry have increased somewhat as the lowered Canadian exchange rate and persistent low-wage strategy (average wage of about $9 per hour in 1997, little more than half the general manufacturing rate) has prompted increased exports to the United States under continental free trade, while the most vulnerable sub-sectors of the industry have continued to receive some protection from duty remission programs (Industry Canada, 2002; Sheikh, 1999). But disrupted forms of production persist as the industry continues to rely heavily on contingent workers shifting back and forth between small movable factories and home-working (see Ng, 2002). The protective duty remission measures will end in 2005 under World Trade Organization rules. The garment industry's general adherence to a low-wage strategy with limited technological innovation and less organized training forebodes further continuing disruption.

Union Survival Response

In the early part of the twentieth century, garment workers' unions were among the largest and most progressive North American industrial unions. As the relative strength of garment workers declined in the post-World War II period, the established unions began to merge to retain their bargaining power and capacity to service their members. In 1976, the Amalgamated Clothing Workers and the Textile Workers Union of America joined forces to become the Amalgamated Clothing and Textile Workers Union (ACTWU). In 1995, the ACTWU initiated a merger with the International Ladies Garment Workers Union (ILGWU) to form the Union of Needle Trades Industrial and Textile Employees (UNITE). The process proved challenging but most of the union offices including those in Ontario were merged by mid-2000. At the time of merger, UNITE represented around 30,000 workers in Canada as a whole with 10,000 in Ontario in over 100 locals. Most of the Ontario workers were ACTWU members, about half of these in garment factories, the remainder in textile plants and a few other related industries (e.g., carpet, fibreglass). The ILGWU component was much smaller, comprising less than 1,000 workers in five locals representing: home-workers, sewing-machine operators and garment warehouse workers, cutters, garment pressers, and social service agencies, retail stores and warehouse workers.

UNITE depends on the central labour bodies,[2] including the Canadian Labour Congress (CLC), the Ontario Federation of Labour (OFL), and projects of local labour councils such as the Metro Labour Education Centre (MLEC) which is supported by the Toronto District Labour Council, for many member support services. MLEC offers Toronto-based garment workers ESL and computer classes. English in the Workplace (EWP) is also offered, usually with the employer's cooperation. Arrangements are made so that the employer pays at least 50 percent of the employee's time spent in class. The UNITE affiliates receive scholarships for their staff from the CLC to attend week-long education programs at the CAW Centre in Port Elgin. Affiliates use CLC training material and rework it specifically for their workers.

Both separately and together after the merger, ACTWU and ILGWU made impressive efforts themselves during the 1990s to provide training and adjustment services to the declining numbers of increasingly fragmented and socially diverse garment workers facing disruptive conditions. The central challenge has been to bring the remaining workers in unionized factories together with the most vulnerable workers in small contracting shops and home-workers. A primary strategy has been community unionism initiatives that link organizing workers with other community organizations (ILGWU, 1995: 63–64)). The ILGWU released its research on the working conditions of home-workers in the Chinese community in 1991 and initiated a series of strategies to mobilize workers in the women's garment sector and to bring about legislative changes to the Employment Standards Act. In 1992, it organized thirty home-workers into the Home-workers' Association (HWA). Many of these workers were previously employed in the King-Spadina garment factory district (Sobel and Stephen, 1996: 19). HWA members now have access to a modest variety of learning opportunities, day-care services and a benefits plan. Funded by a variety of sources, the HWA is a pre-union organization that operates as a community-based organization. It has developed a variety of programs and services for its members, including: legal and social services; recreational activities; workshops on workers' rights, health and safety, parenting, and setting the piece rate; and ESL, sewing and leadership classes.[3] Members are slowly introduced to the idea of a union as they become more involved with the Association's social and learning activities. As of 1 June, 1996, UNITE began offering home-workers and unemployed workers computer classes, sewing skills classes and leadership training. Garment workers who are landed immigrants can register for government-funded English classes which are offered on weekends at the union location.

The ILGWU reached out to its diverse membership by translating its newsletter into several languages: Punjabi, Cantonese, Spanish, Italian and Tamil. The UNITE regular newsletter is offered only in English and French. The hiring of a new UNITE Education Director in the mid-1990s led to the development of a number of other new educational initiatives.

In addition to standard grievance training, courses include harassment training, conflict resolution training, and leadership training. There is now a strong push for stewards to better understand how their role affects their workplace and their community, and to learn about the national and international levels of the union. Courses that were more localized around workplace issues are now being expanded to examine the ways in which the local affects how the national and international unions work. There is more encouragement for more political and social activity in the community.

Due to the large number of plant closures since 1988 and the large number of displaced garment workers, labour adjustment strategies are crucial in assisting garment workers in returning to the labour market. ILGWU/UNITE had two programs for displaced workers which were funded by the Ontario Training and Adjustment Board (OTAB). The York Action Help Centre met the short-term needs of 200 unemployed workers and provided them with a range of labour adjustment services. The Garment Workers Action and Adjustment Team met the long-term needs of about 400 workers and comprised a multilingual staff. Displaced workers were offered sewing classes to upgrade their sewing skills. Both of these programs were disbanded by the provincial Conservative government in December 1996 and March 1997 respectively. Nearly half of currently unionized garment workers, who are over age forty-five and face the most extreme forms of potential displacement,[4] were disqualified from receiving training and education funding from the provincial Transitions Program.

UNITE has continued to develop a more proactive training and adjustment strategy. As the major ILGWU report (1995: 66-67) stated:

> The guiding principles behind the sectoral adjustment strategy include: meeting the immediate needs of immigrant women such as language, timing, location of meetings and the provision of child care. The new program hopes to address the needs of the long-term unemployed. At the same time, the union can take a more active role on training issues in the workplace. The union needs to intensify its efforts to provide ESL and other upgrading for workers.... The union should provide a union-based educational program on technological change to workers and create the opportunity to debate issues of technological change and work re-organization.

But this report also recognizes that the union faces a seemingly "overwhelming task" (69) in organizing in the face of such disruptive conditions, especially for home-workers and those in smaller enterprises.

Profile of the Workforce

According to the most reliable estimates, women comprise over 75 percent of the garment industry labour force compared with less than 25 percent for manufacturing generally. The workforce is highly segregated with men dominating the cutting room and women filling most of the sewing-machine operator posts (Grant, 1992). The industry has long relied on immigrant labour, especially women, as an inexpensive labour pool. The ethnic composition has reflected the flows of "immigrant"[5] workers. Historically, English literacy skills were not required so the garment industry has served as the first point of entry into a gendered and occupationally segregated labour market for many new immigrants. The industry attracted many Portuguese, Italian, Greek, Macedonian and Polish workers during the 1950s and 1960s, while Vietnamese, Chinese, South Asian, Caribbean and Arab workers represent more recent flows.[6] The garment workforce also tends to be older and less well-educated than industrial workers generally (Grant, 1992). As one key informant observed:

> The apparel industry was built on immigration. Whatever the immigration flow into Canada, you can track it in our industry. They are entry level jobs, very little language has been needed, [formal] education skills have not been high... The diversity is really huge. Dozens of ethnic groups in one plant. It changes from month to month. We get them as they land here and it's first entry jobs. Language has not been required. But now they're expecting more skills, grade 12 in some factories – just because of the high unemployment.

The labour-intensive character of the industry is indicated by the large share of production workers in the labour force, nearly 90 percent compared to around 75 percent for manufacturing generally, and by the much higher expenditure on wages in relation to the value of shipments – even though wages are very low in this industry. Adaptive low-wage strategies have therefore driven employers in the garment sector to rely on the external skill formation of their workers and to invest little money on retraining their internal labour force. Thus, workers who seek employment in the sector have often already acquired most of their relevant skills informally in their native country and, in the case of women, through learning to sew at home.

In the wake of trade liberalization policies and foreign competition, the domestic garment industry has become precariously dependent on these adaptive low-wage strategies. In 1996–97 when most of our interviews were conducted, garment employers were increasingly drawing a contingent labour force from large numbers of unemployed factory workers, a steady flow of non-English speaking immigrants from the Third World and Eastern Europe, and a marginalized female labour force. As one of our interviewees described the recruitment process for the small garment factories:

> They're choosing people. They came to the Vietnamese Association, picked four ladies and told them that now they had to work. If you work faster you'll be chosen. Four worked very hard but they chose two and they threw out the other two. The next day, they came back to the Vietnamese Association and got four other ones and they chose another two. My sister only worked three days because they said she was too slow. My sister she knew she has to earn money for herself so she works hard.

Our study of garment workers began in the summer of 1995 with the initial assistance of ILGWU/UNITE. ACTWU/UNITE subsequently provided access to additional members working in larger factories. Union staff members provided the project with key background information. Our sample is exploratory and consists of representatives from the three segments into which the garment workforce is increasingly being divided: employed factory workers, unemployed unionized garment workers and home-workers. Twenty-four interviewees participated in the research, including fourteen employed factory workers, seven home-workers and five unemployed workers. Only unionized workers and home-workers associated with the HWA were included. The most extensive profile of the garment workforce available to us when selecting the sample was a survey conducted by the ILGWU/UNITE (1995) of its factory-based members. Table 7.1 provides comparative demographic profiles of the ILGWU survey and our sample.

The majority of the interviewees in both studies were female, reflecting the well-established gender composition of the industry. The interviewees' ethno-linguistic backgrounds were fairly similar, with Chinese dialects being the most common and Italian and combined South Asian ancestries (including Vietnamese, Punjabi and Tamil) being the next most frequent. Educational levels are also fairly consistent in both samples with some or complete high school being most common while college-level schooling is quite rare. In both samples, the remaining garment workers tend to have a long history of employment in the sector with the largest concentration being more than twenty years of experience in the sector. The occupational distribution is also quite comparable with sewers comprising the largest group. The segregated division of labour within the garment sector is further reflected in our sample by the fact that the female respondents were predominantly sewing-machine operators while the male respondents were nearly all cutters. We make no claim to having a representative sample, given the very temporary nature of many of the work sites. We have tried to include a wide range of garment workers. According to the ILGWU/UNITE survey, nearly 20 percent of factory workers were unable to understand work instructions given in English and so our research team conducted several interviews in Cantonese and Punjabi.

Table 7.1 Demographic Profiles of Garment Workers

	ILGWU/UNITE Survey Profile, Unionized Workers, 1995	WCLS Survey Profile All Workers, 1996–97 (N=24)
Gender: female	80%	81%
male	20%	19%
Age: 20–30	11%	13%
30–40	30%	52%
40–50	27%	17%
50–60	30%	22%
Educational Attainment:		
Elementary School	30%	25%
High School	63%	53%
Univ./College	8%	12%
Ethnic Breakdown:		
Chinese	30%	45%
South Asian	20%	20%
Italian	20%	13%
Other European	30%	8%
Other	—	8%
Seniority at Work:		
Less than 5 years	13%	21%
6 to 10 years	21%	8%
11 to 15 years	10%	21%
16 to 20 years	20%	8%
over 20 years	31%	41%
Occupation:		
Sewers	49%	54%*
Cutters, Finishers	27%	25%
Packers, Shippers	14%	12%
Others	9%	8%

*includes seven home-workers

Our research included alternative sites such as community organizations and language training programs to allow a more inclusive sample of learning processes among the garment workforce but it is likely that non-unionized factory workers and more isolated home-workers may experience even more difficult working conditions and further barriers to education and training. Our respondents were interviewed in various settings: some in their homes, others in a familiar organizational environment, and some in focus group formats. Their accounts of their employment experiences and learning activities make up the rest of this chapter.

Job Disruption and Stress

Our in-depth interviews with garment workers revealed a great deal of stress and uncertainty over changing working conditions as well as workers' general confidence in their technical competence to perform the tasks required in their jobs.

The women sewing-machine operators all talk about speed-up and constant change in production requirements. Paula is an Italian immigrant who works for a men's sportswear manufacturer as a serging machine operator. Her work is monitored by a computerized system that is pre-programd to allocate the material to the various sewing-machine operators via a mechanized distribution system. The computer monitors the workers' efficiency and production rates and time-and-motion analysts are also used. Paula works with twenty to forty different types of material and has a situated knowledge of 100 different styles. She is a highly skilled sewing-machine operator and has been working in the garment sector for more than ten years. She must concentrate constantly because, if she is distracted, she can easily cut the material and ruin the garment. Her major concern is production speed:

> You need to pick up speed, like you know, to produce more. When I started maybe I was 80 percent, maybe now 110 percent. You have to go more. A guy comes, he times to see how many minutes you need to make that operation. After that, for 100 percent you have to do like 15 units in one hour. So I try, try, try.

Joya is a Portuguese sewing-machine operator who does piece work on a mechanized sewing line making diverse styles of sports wear. She complains frequently that she has little time for anything but her job, commuting and taking care of her family. She is also concerned about speed-up:

> It comes cut on hangers and we sew and then it goes up again. It's on computer now, a few years ago it used to be bundles on top of the machine. The actual sewing hasn't changed, just the way the work comes to

the operator ... Sometimes I get upset because they put the rates down, oh, it kills me. The last manager put the rates down a lot. We've been trying to do it and we can't do it. We can only do so much.

Donna, a Caribbean immigrant, works for a men's garment manufacturer and has more than ten years' experience in the garment industry. She knows how to operate both the single needle and serging sewing machines. Her current employment requires the use of the single needle sewing machine and her assigned operations are sewing the front and back pockets on a pair of pants. However, she also knows other sewing operations and can be called upon to back up other sewing-machine operators. Her extensive knowledge of different fabrics and how to work with them was developed informally over years of experience. Donna finds some security in unionized factory piece work compared to non-unionized garment sewing work because:

In piece work it sometimes goes up and down. If you're doing cotton, linen, okay, you might earn some money in that. But it's the type of polyester viscose that's got any lacquer in it, it would stretch a bit so you lose time on that. Piece work goes up and down. But it balances off.... When I came to work in this factory it was much easier [than prior non-unionized garment work]. You just sit down and you just serge pockets and at four o'clock you go home and you forget about it. I don't know if it's because I like to sew or because it's the only decent job you're going to get. You always have the strain but you must like something about it.

The packers in our study, as in the industry generally, started out as sewers but were not fast enough. For example, Pramila was given a few months on a sewing machine after she arrived from India but was soon put in the distribution unit of a women's clothing company. She has worked there for over five years while most of the production workforce has been laid off and the company has come to rely almost exclusively on contractors to make the clothing. She stresses how much the owners like her work because she is a hard worker. She also describes an intensified work process:

The garment items are organized by style with their corresponding sizes. Each style of garment has a separate reel. We receive the orders from the retailers by "Style Number," "Size" and "Number of Pieces." We proceed to fill out the order on a special metal hooks: We put bags on it, and tie them. Then we put stickers on which store they're supposed to go to. The trucks come every day. There are only a few of us doing all the packing now.

The distinction between unemployed women garment workers and home-workers is sometimes arbitrary as unemployed workers resort increasingly to home-working and home-workers seek factory work. For example, Hoa is a Vietnamese immigrant who formerly worked as a sewing-machine operator. She has had a series of temporary larger factory, small contractor and home-working experiences since arriving in Canada nearly a decade ago from Vietnam. She sees no attraction in home-working:

> I don't have anything, just the same as when I came here. I'm hoping to find a job to feed myself and my children – that's all. Many women are working from their homes and being paid very low wages because that's how they're living. We want to work for a big company where they pay tax and get benefits. The lady I've been working for recently would not talk to you because she's doing contracting work out of her home with four machines in a small room and not paying people right.

Hoa tried home-working after being laid-off, but subsequently found another factory job and would not go back to home-working. The time crunch was just to great for her:

> I quit home-working because I cannot manage the time. I worked a lot of hours, sometimes from 8 A.M. to 2 A.M. Changing a lot, sometimes I work a lot and sometimes nothing to work. I don't like that. I cannot control the time. When the company cut the garment, they phone me to come up to get the work and I don't know how long for this work. I work until it's finished. Sometimes the children help me to cut the thread. I don't like that but I have no choice because the owner always push to finish for him.... I would not go back because I cannot control the time. In the busy season about twenty continuous days [of work], I got a lot of stress. Very difficult. The time. The most important thing in my life.

Hong is another Vietnamese immigrant who continues to do home-working in cooperation with some friends. Through a translator, she also expressed the increasing desperation involved in this work, which she does on a very old sewing machine in her basement apartment:

> We get paid $1.50 a piece and make about four pieces an hour. We have to drive to another city to pick up the material. It costs about $15 for gas. We get this work through a friend in the factory, which is better than going through a jobber who takes half the money. We continue to work from home because we don't speak English and it's the only work we can get. Somebody in the community knows where to get material from and they tell us. This time it's very hard because the price is getting cheaper and cheaper and there are plenty of people to work.

The male cutters generally expressed much concern over factory downsizing and computerization as job threats. Jerry's versatile cutting and marking skills were "highly valued" in the Canadian garment industry when he arrived from China around twenty years ago. He was thus able to secure good employment as soon as he arrived. But times have changed:

> I've been happy at most of the companies I've worked at because they were all unionized with job security, regular raises and benefits. Now there are non-unionized factories and subcontractors everywhere. Unionized factories are closing down because they can't compete with the lower prices. Many workers are just thankful that they don't get laid off. So, they will do work that is not part of their job. If a cutting job is being phased out by computers, the cutter will have to do ironing if the boss tells him to. What can he do?.... So now, even unionized skilled workers do not have security.... Also, many factories are very unsafe, they pile up the clothing so high that they often collapse.

Tony came from Italy as a fully qualified tailor about twenty years ago. He waxes nostalgic about skilled tradesmen's recent loss of control over their jobs:

> The company where I was before, they called me. And the manager he says, 'We need you here. We give you more money. We give you what you want'.... I was allowed to talk to the design person of this company, the owner. So I was the one that could communicate straight on the design, like whatever I want. Can we still do something like that, you know? ... Then they bought a computer to do the work. I said, 'I've got nothing to do here.' The manager said, 'Just watch the machine.'

David is an older and even more experienced cutter who became a tailor in the Middle East and has worked in Canadian factories for over twenty-five years. He offers essentially the same story:

> I'm currently using a round knife machine or a straight knife machine in a women's coat factory. Been there one year. I've been laid off and called back. The new technology means less work and less pay. I trained on the new computerized equipment when I was laid off. But they've hired someone else to do it. I figure that's because they pay the new person less than they would have to pay me. Now it's hard to find any job at all.

Together, these diverse accounts provide a complex story of skill development, struggle and differences across a wide range of workers in the garment industry. They provide the basis for our analysis of work and learning in a sector in which basic patterns have been severely disrupted.

Organization of Garment Sector Training

In line with the generally adaptive managerial strategy of employers, the provision of training programs to prepare for widespread workplace changes has been quite minimal. A survey of union representatives in 1994 found that three-quarters reported no increase in employer-provided training in their workplaces over the past five years, and 60 percent felt that their employers offered too little training (UNITE, 1995). The general conclusion of a union-sponsored research project to evaluate existing training and adjustment programs in the mid-1990s was that:

> In the face of a fundamental restructuring in the garment industry, it is the union and workers themselves who are left to pick up the pieces. While the federal government has promised to provide an adjustment program, no coherent plan was ever developed. The existing adjustment process is disconnected from any sectoral plan. Most often, employers are unwilling to invest in workplace training. Workplace training initiatives continue to be on an ad hoc, firm-by-firm basis. (ILGWU, 1995: 66)

The development of sectoral approaches to tackling the training and adjustment problems facing the garment industry has been very slow. Only a minority of workers are now unionized, employer opposition to unions in many workplaces is quite strong and the industry's central bodies have shown little interest in working with labour. In 1993, the Canadian Apparel Federation (CAF) was established to implement a new industry-wide strategy to address issues relating to the image of the industry, labour-management sponsorships, education, training, re-training and development. Unfortunately, the strategy focus in the first report excludes any mention of skills training for operators (Fashion Apparel Sector Campaign, 1991). Manufacturers' common strategy of driving down labour costs by using contingent workers isolates home-workers, undermines unionized plant workers' job security and limits prospects for any labour-management partnership on training matters.

The Fashion Industry Liaison Committee (FILC) of the City of Toronto, which had representation from manufacturers, labour unions, educational institutions, and government officials discussed the recommendation from the 1994 Apparel Textile Action Committee (ATAC) Final Report (Ng, 1994) to create a training centre for the garment sector. There have been more recent efforts to conduct government-funded joint labour-management assessments of the industry's human resource needs, including the Apparel Human Resource Council created in 1997 which includes UNITE and a few other unions that have members in the industry. But these efforts face the same limits as earlier ones. The early 1990s survey of garment factory workers found that only about one third of employers offered training on newly introduced equipment (ILGWU, 1995). It also found that about three-quarters of factory workers felt they had skills that were not used on a daily basis. An elected plant representative may be expressing a quite common sentiment when he states that:

> It's not difficult to learn new things but the employers are not patient. They want you to speed up right away. They expect you to produce twice as much yet they don't provide any training. They want double production the same day. (ABC Canada, 1996: 3)

A Profile of Learning in the Garment Sector

In the context of these disrupted and stressful working conditions, garment workers told us about a wide array of job-related learning activities, many of which were done informally and visible only to themselves.

Formal Schooling and Job Training

The lack of recognition of educational credentials from non-Western societies has been quite well documented (e.g., Livingstone, 1999a). Immigrant workers now usually arrive without English-language skills, so those who find employment in the garment industry are particularly likely to find any formal education they have to be undervalued. This appears to be true even if they have relatively little schooling.

Most of the female garment workers we interviewed had formal schooling careers that ended early. Suni, for example, left school in Hong Kong at age twelve to work illegally as a factory sewing-machine operator to help support the family. She regrets that she could not continue her studies because there was no one to support her at home. Her brother was favoured in his studies and Suni had to help her mother with household chores until she was too tired to finish her homework. She took a course at a special dressmaking school to gain more comprehensive knowledge about the business before coming to Canada but for the most part, she has gained any new skills she needs by watching other skilled operators. Joya managed to complete grade 8 after she came to Canada. She hated school mainly because her lack of English made it difficult. She now regrets this a lot but has similarly felt no need to take further courses related to her job:

> You don't need much formal education to do my job. To me it's fast to learn. You don't have to have grade 8 or grade 9 to learn how to work. As long as they teach you for three or four days you learn fast. I did know a little bit about the sewing machine from home economics courses in

school when I started working at the factory. But there's no other formal training for my job.

Donna has what she calls "a good high school education" in the Caribbean. She also took a sewing course before she finished school. But she has no other formal job training.

> I took a sewing course while I was still in school before I came to Canada. I took a dressmaking course to learn to sew my own clothes. I never thought I would end up working in a factory or anything like that. I just took it to learn to sew my own clothes. I learned basic sewing at school but I learned more at home with my mom fixing my school uniform. We got an automatic machine shortly after I started to learn at home and then it was a 'breeze.'

The women packers and home-workers with working-class origins gave similar accounts of leaving school early, lack of teacher support and the need to work to help support their families in their native countries. But they indicated unanimously that there was no formal job training required for their work in the garment industry.

However, in general, female garment workers' views of the value of formal education were positive. As Maria, an unemployed Italian factory worker puts it, "It's important to have an education anyway so you can use it when you need it. It's important to know things, even if you can't use them." But educational attainments varied depending on class origins. Women from poor working-class backgrounds were discouraged from pursuing their formal education past the primary level prior to employment in their home countries. Those from middle-class backgrounds had a secondary or college education and tended to be underemployed in the garment sector because of the lack of accreditation process in the Canadian educational system. None of the employed respondents participated in continuing educational programs offered by, for example, community colleges. Home-workers who are members of the garment union's Home-workers' Associated have access to specially designed weekend classes in sewing, English and leadership skills. Among the unemployed women workers, only a few were in retraining courses through Unemployment Insurance.

The male participants are all middle-aged and have all been employed as cutters at one point in their work careers. All of them served extensive apprenticeships after leaving school at the end of high school or earlier, typically in their native countries. David left school at fourteen after running away from school a lot because teachers were violent. He disliked school but now enjoys studying. He apprenticed with his father who had a tailor shop. He has regularly taken courses since coming to Canada, first language courses, then numerous upgrading courses, especially on computerized equipment. He talks of the benefits of taking courses both in terms of enjoyment and amassing diversified skills for other jobs. Tony also had a long apprenticeship in Italy after leaving school at sixteen:

> It took me eight years. It's not like here you go to school. You learn in trade school, you go out, they use you. Like they make you do house chores. You have to start by some days [sic] just pulling thread. And then if you're good, they will give you extra work, like now you stitch the side seam. After that, then you do the pockets. It takes years before you get to the pants, to the jacket. You don't become a tailor just because you want to. I did this for eight years and then I became a tailor.... I never pressed before but I went into the press, looked up the guide and did it, it doesn't need a formal education. I'm a tailor. Now you tell me that you have to teach me to use the sewing machine? Pah! Give me a break! I can do it. Just let me look at the machine.

Jerry finished high school in Hong Kong and went directly into the cutting trade as an apprentice in a factory. He,

> started by pulling cloth. After a while, I learned how to cut facings, and then cutting linings, before I was able to cut the fabric. Because I had high school education, and was a quick learner, I was given the job as a 'marker' soon after I started. My job does not require higher education, but does need grade 8 education in order to understand what needs to be done.

Jerry is now actively engaged in retraining courses to find another job but is not optimistic about either his chances or employer support for older skilled tradesmen inside the industry:

> Now that I am getting older, I want to retrain for something that is not as physically demanding. But I have to recognize my own ability. I would like to learn computing but I realize that my English is not good enough. So, I'm retraining in [another marketable skill]. I am not confident that I will be able to complete the course. But I will try my best ... In terms of retraining in the industry, unionized factories have provisions for retraining in computer technology, etc. Even so, they will lay off most and only retrain a few. They would only train those who are young, smart and have initiative. Non-unionized plants will simply hire already skilled people. They wouldn't waste their time and money on retraining their workers.

Despite their limited initial formal schooling, most of the men were participating in various upgrading or retraining activities and faced fewer barriers than women accessing these programs. Unionized cutters were offered retraining in computer technology as part of the negotiated contract with the employer. Perceptions of the impact of computerized technology and their future job prospects played a more direct role in influencing cutters' choice of work-related learning activities than their prior schooling. For example, even though David has only a primary school education, he has participated in extensive computer-related formal training in preparation for the garment sector's switch to computerizing the cutting trade. In contrast, Lee, who also has a primary school education, did not think it was worth his while to pursue computerized training because he did not think he would be hired because of his age and seniority at work.

Age is indeed an important factor that generally influences garment workers' interest in participation in further job training courses. Younger unemployed workers whom we interviewed were eager to change occupations and were willing to be retrained in a different career. Older unemployed women workers often felt they had to rely on their job experience to try to find other garment work. One of the male garment workers who is currently retraining at a community college finds it difficult to keep up with the schoolwork because of his age. In contrast, a younger respondent who is also retraining is able to keep up with both the English requirements and the schoolwork.

Informal Job Learning

Women garment workers acquire almost all of their job skills informally either from their own observation and practice or from other workers or supervisors. A focus group with home-workers who had all previously worked in garment factories offers a clear account of this process:

> We all learned to pick up skill on the job early. We got no formal training. We started doing general labour tasks like string cutting, cleaning and packing. Through "secretly" observing other sewing-machine operators, we picked up skills kind of unsystematically. We'd practise during lunch or tea breaks. Then we applied for other garment work and from there learned more skills on the job, either from other workers or from supervisors. We gradually picked up more skill on the job until we had experience in all work procedures and could finish a whole piece of clothing or dresses of different designs by ourselves. Having good supervisors was critical to learning all kinds of skills.

The current factory sewing-machine operators tell similar stories of informal on-the-job training. In Maria's words:

> I got the job through a friend who worked there. I started off working at the cutting table, cutting thread because I had no machine experience.

214

After a year or two I learned how to use the button hole machine. The supervisor taught me how to use it.... Experience on the job has been very important for me and the people I know.

Donna entered the factory with some sewing skills, worked immediately as an operator and continued to learn mostly on her own.

To be honest, learning to work with different fabrics comes down to your own skill, no one teaches you how to handle the fabric. I worked a lot as a 'float' – if someone was away, I would go and do different jobs. That's something I taught myself. I looked to see how they were doing it and just pick it up. The first place I worked the foreman would just stitch one and then they [sic] got up and you had to know how to do it ... I like to sew and if I can't figure it out I'll come home and think about it.

But she does feel that factories could be better organized if they gave people more organized job training and especially recognized the difficulties created by language barriers.

Don't just dump them in the corner because they cannot speak English. Right now they just show them how to use the machine. You see people sometimes, they can't do something and they cry and it bugs me.... Now workers have a shorter training period and then begin working as piece workers. Three months is often too short because the workers are moved from one task to another during that time so they don't have the opportunity to develop their skills, build up speed. If they did it straight for three months it would probably be okay.

Most of the men learned the cutting trade through an apprenticeship program in their country of birth and often enhanced their job-related skills through collective informal learning in various garment factories. As Jerry recounts his apprentice experiences:

In Chinese society, to do my kind of work, one is required to be an apprentice. But it is up to you to learn from others as well. You need to be observant, and ask others questions, and then follow how others do it. I learned it in Hong Kong. I did not have a teacher. I learned everything on my own. I did it out of necessity, I didn't have a choice.

Lee was enrolled in an apprenticeship program, but his "master" exploited him and did not formally teach him the cutting trade. He learned his trade informally by "stealing" from his master and by informal group learning with his fellow apprentices with whom he shared a room. He learned a lot of skills from the other apprentices. They taught each other how to make buttonholes, make collars, sew side seams, etc. The "master" did not teach him everything, but he "stole" his master's knowledge by watching him and imitating what he did. Sometimes he would invite another apprentice who knew some other skills for tea, and then ask to be taught the skills. The apprentices helped each other out.

The male cutters who provide such detailed accounts of their formal and informal apprenticeship training offer relatively little information about continuing informal job-related learning as cutters. David discussed learning different techniques when moving from sportswear to coats, in particular using glue instead of weights for cutting the material and learning some related computer techniques on his own in conjunction with the courses he has been taking to prepare for the introduction of new equipment. Tony describes his curiosity to keep learning new methods and the lack of encouragement he generally received from the company:

> In the company, the more you want to know something, the more they put you down, it's not your job. They don't give you a chance, they call somebody from outside. But I never wanted to sit at my table and do the same thing. I used to go and find something even if it wasn't my job. I had to go and move.

But generally the skilled tradesmen tended to regard their formative apprenticeships as central to most of the skills they have subsequently used, to consider that most of their continued learning has occurred through training courses and to speak of relatively little continuing informal job training.

These accounts should not be taken to suggest that employers are generally indifferent to workers' training, but rather to confirm that they are heavily reliant on workers' own efforts to train themselves. In addition, "cross-training" is increasingly encouraged. Employers now encourage their workers to take on several operations in order to "keep the line moving" when workers are absent. This is especially important for them during low production periods. A worker who can be deployed as a "float" in a factory enhances his/her chances for remaining employed. Lee is employed as a cutter but he also does "a lot of odd jobs as well ... whatever needs to be done, I'll do it." Donna is trained not only to perform her own piece work sewing operation but several others' as well.

Female garment workers who are employed as home-workers undertake an enormous amount of job-related informal learning. Many of the home-workers obtained their skills by working in garment factories. One home-worker learned her skills when she worked with another experienced home-worker. Skilled home-workers invest in their own industrial machines. They learn how to operate and repair their machines, how to handle a particular fabric with the machine, and the type of stitch to use for a specific fabric. With each new order, home-workers learn a new pattern and design. They are given a sample to work from but, working isolation, they are ultimately responsible for their own training.

Skilled sewing-machine operators also have a vast range of skills and work-related knowledge on which to draw. These are gained on-the-job, by switching from factory to factory and learning as many operations as possible. Skilled operators also sew in their homes, either a hobby, for their children's garments, or as extra cash income by performing alterations. They learn while they are sewing at home, working with different fabrics and using various patterns. The employer draws from this knowledge since he does not offer any systematic on-the-job training but the operators are using their knowledge acquired during their own time for performing their jobs as skilled workers.

Even though cutters are the smallest group of workers in the garment sector, they were the only job category that accessed formal job training in the form of computerized technical training. The interviewees who worked as cutters continued to apply their trade knowledge to their work but were also expected to perform other, often less skilled, responsibilities. Indeed, virtually all garment factory workers were expected to take on an increasing number of tasks to maintain the production cycle under and heightened competition and widely fluctuating demand.

Union-Based Learning

While union leaders have recognized the pivotal role that training and adjustment programs should play in sustaining the garment industry, relatively few workers have participated in union-education programs because of the economic crisis in the industry and persistent barriers, most notably language limits and lack of time among female garment workers. Those in our sample who participated have generally found the experience to be quite helpful.

Donna and Pramila participated in union-organized education programs related to their roles as elected local leaders. Both have experienced enlightening leadership courses like Negotiating the Contract, Leadership Training and Shop Steward Training. Pramila has also had extensive exposure to a Labour Studies curriculum at the CAW Port Elgin Centre. Courses included labour history, sociology, economics, political science and labour law. She describes the experience as both generally very informative and deeply sensitizing to previously ignored issues of racism and sexism:

> It was *almost* perfect. It included all kinds of valuable information and being there was like being with my own family. I was impressed with

hearing about the early Indian immigrants to Canada and the difficulties they faced.... But in my workplace, Indian, Chinese, Jamaican, whatever – we all stay together like friends, right? So when I went to [the study program], I really was feeling like [white] people giving me discrimination. I was feeling really, really worse myself, right? Sometime [sic] I started crying myself, because I was feeling "why I come?" I shouldn't have come here. I chose the wrong place. But then I have to fight for this struggle. So I spend my time, but I feel very much isolated, like it was like racism I feel ... But the information has helped me a lot to see and fight more forms of discrimination.... I also started to question more why women at work were being beaten by their husbands, which is totally wrong. We really need to get more union education programs into the workplace.

Members of the HWA who have been organized around specific weekend courses say they also greatly benefit from being in contact with one another, thus breaking the isolation of their separate workplaces. Sharing information over the telephone or at union meetings is one of the ways members learn collectively.

Jerry learned his union work informally while volunteering as a union activist. Through his union involvement, he became more interested in politics. Labour studies courses helped him to better understand his own experience of being exploited as a worker. He indicated a strong commitment to working through union activities to fight for a just society where wealth will be shared more equitably, as well as a sober sense of the limits of individual struggle:

I've learned a lot. Learned how to organize workers, the advantages of forming unions, being able to voice our rights, and being able to ask for health and safety standards.... But it is important that the people be united, and that they trust their leaders and understand that they are exploited by their employers. Sometimes, if the employer really wants to fight the union, they would go to the extent of closing down the factory.... It saps your energy completely and your health suffers.

English-Language Skills

Like all the others we interviewed, garment workers also told us about a wide array of other learning activities not directly related to their jobs (see Chapter 8). Their interests in diverse areas of general informal learning appear to be limited most severely by female workers' lack of time to pursue them. Few workers participated in courses not directly related to their current or prospective employment. The very important exception, universally seen as vital for both job success and life generally, was English-language learning which was actively pursued both through courses and informally.

Almost all workers talked about the need to improve their English skills to qualify for better jobs and about the challenges they faced to becoming fluent. Most have made persistent efforts through whatever courses were available to them as well as continuing informal learning to improve their language skills, both for their employment prospects and to help their children. Mei offers her account:

> Mistakes I made after coming to Canada made me realize how much of a problem it was to be ignorant and dependent on others, and how important it was to pick up more life skills, especially language skills. When my son was two, I began to learn English systematically by attending classes in public libraries where childcare was provided. I did home-working and had no more time to go to English classes. I bought tapes recommended by ESL teachers and began self-study at home. I practised with tapes and did exercises when I had lunch breaks from home-working.

Hoa has a similar story:

> I need to understand things. I need to know what is happening around me. When I ride the bus, I see people read the newspaper. I am very eager to know what happened. I also have to know English to understand my children. I took English at an adult learning centre which was very helpful for me.... I regularly watch the news programs. I think it's the fastest way to learn English. I listen to the radio when I work. Sometimes I hear other people talking and I learn from their talking. I learn everywhere, even the advertisements in the subway. When I see vocabulary I don't know, I go home to look it up in the dictionary.

Employers do not appear to view English literacy as a necessary skill for sewing-machine operators but from a worker's perspective, it is essential for optimal job performance. Donna, one of the most fluent workers in English, describes how she applies her reading and writing skills to her work, both as a local president and as a sewing-machine operator:

> [As Local President], because most times, sometimes, they tell you something and you got to put down the date, you got to record when it happened, how it happened, everything you got to take reports of everything, you understand.... Also, for the sewing for my job it's very important, too. Because sometimes I put on the labels. You have to read the labels to know the fabric. You have to know the fabric, like, this is polyester viscose, you have to know if it's wool, if it's linen, you have to know the fabric, that is very important.... I can also talk to my supervisor, I can understand what they are talking, what they are saying, right?

And Jerry, like all the cutters who took English language courses soon after arrival, continually worries that his English will still not be good enough to change careers:

> My English is really not good enough to learn advanced computing skills. Maybe the course instructor didn't know that I don't have sufficient English language proficiency. But now I have to do a lot of written tests and it is very difficult for me. For the kind of work I want to do, I really, really need better English.

In general, men had more access to organized English classes when they first arrived in Canada and, therefore, greater opportunity to pursue their learning for several consecutive years. Women, on the other hand, were impeded because of childcare and other home responsibilities, as well as policy criteria limiting access to "heads of households." The female respondents' English-learning strategies did vary somewhat: older workers learned the language by "picking it up" and did not attend formal classes while others took English classes for a few months when they first arrived in Canada and later continued learning informally from co-workers, friends and family. All the female garment workers faced more barriers than the men in accessing organized ESL and basic literacy skills programs. Their lower basic English literacy skills compound the constraints they confront in pursuing opportunities for job improvement.

Barriers to Learning for Garment Workers

Workers from minority ethnic groups generally face more barriers than the dominant ethnic groups when accessing organized learning opportunities. Their minority status, need of ESL and limited financial resources are some of the basic constraints, similar to the situation we explored in the previous chapter. Female garment workers encounter more barriers to learning because of their gender roles and lack of discretionary time. Male garment workers experience barriers to learning because of a lack of English skills and, in some cases, age. It is important to note that there is significant overlap in how the barriers impact on individual workers.

There are striking differences in learning activities by gender. The male garment workers accessed various formal training opportunities related to their jobs, starting with extensive apprenticeship programs in their skilled trades and including union-negotiated adjustment programs. They faced fewer barriers than women to participating in upgrading programs. Women had few formal training chances, relied almost exclusively on informal job-related training, whether on the job or in the home from their mothers and friends, and qualified less frequently for adjustment programs. The greater underemployment of women's formal education in the industry is not only exemplified by the women with college degrees who are home-workers. Sina, for example, is a fully qualified electrician who can only find work as a sewing-machine operator. Paula, a sewing-machine operator and the local union rep, summed up the situation well when we complimented her on her impressive serging of a whole garment. She laughed and said: "I know. I really am a skilled sewer. I've mastered my trade and the knowledge of sewing techniques. If I belonged to the social group of men, I'd be getting paid double my wages!"

Men described enjoyable engagement in informally learning a variety of recreational activities. Women garment workers, on the other hand, did not indulge in many leisure activities. The differences are attributed to women's primary role in the home as caregivers and mothers. These women's discretionary time is severely constrained since most of their time is taken up with work, whether it is wage work, housework or mothering work. The persistence of sexist attitudes among husbands, employers and workers themselves contributes to the problem. In Pramila's terms, this subordinated role needs to be seriously challenged for women to have fair opportunities and to participate in deliberate learning activities.

> Indian women have oppression from their husbands, from in-laws – they have to go which way they are told and they're not allowed to participate in any activities. Their husbands beat them, still they couldn't speak out for that. I really like to help them.

Lack of discretionary time is a closely related barrier to learning for women. As Donna observes:

> You have to come home and you have to do everything. And then they come home, they sit and look at sports or they take up the newspaper. Sometimes you come home and you are very, very tired and you have to cook, you have to clean, you have to do everything…. It's too much of a strain on you and that person is not doing nothing. And you can't stay home because you need two pay cheques.

So women who work in a garment factory or as home-workers are physically drained and can find neither the time nor the energy to even think of formal courses. As Kim complains:

> Because I work in a factory I know worker feeling. They don't have time to read. They read books for entertainment. They don't have time to think, to have this thinking, after they forgot everything… maybe the work is so heavy so hard they need the bed to lie down much more than the book to read. For them reading [is] the dessert not the supper.

Nearly all the respondents needed further ESL training and wanted to achieve English proficiency. Men were at least able to pursue English courses for a sustained period of time. Rarely did the women have this opportunity. Lack of ESL training seriously constrains women's formal training opportunities and thus limits their ability to improve their lives.

Financial barriers are also significant factors that affect participation in part-time or full-time organized courses:

> Most immigrant garment workers really need money. I mean, they don't have their own house; they live in rental places; and they have to send their kids to school. They think about the future for their kids, right? So they have no time to go to full-time school, to university and whatever, right? So, they're just looking for the job where they can get money to make everything easy.

Nearly all respondents find themselves marginalized in a racist society (Armstrong and Armstrong, 1995; Henry,1994). Because of their minority status, they experience barriers to learning that are specifically attributed to their racial and ethnic origins and their lack of English skills. Minority workers seek assistance from their local community networks and shy away from becoming involved with institutions that tend to represent the dominant group. Jerry believes that many immigrant workers are hesitant to unionize because of fear of losing their much-needed jobs. In this way, he suggests they are barred from learning opportunities that union membership offers. Hien, who is sponsored by her sister, is not allowed to seek governmental financial assistance since immigration policy stipulates that sponsored immigrants have to be supported by the sponsoree. Her sister cannot afford to financially support her so Hien has to find employment. Her main concern is finding stable employment so she can feed her son and afford her rent. She has neither the time nor the peace of mind to learn English at the moment. It is Hien's material reality that dictates her life situation:

> I have one son, an apartment. Right now I am working and make around $800 a month. With rent of $580 and the telephone over $20, I have $200 left for the bus, for food, for everything. That's why I am thinking of every cent I spend. So if they have any class around me, I am willing to go. I need anything so I can work.

Older workers mentioned their age as a significant barrier to work and learning opportunities. David believes that despite attempts at retraining, younger workers will always be preferred over older ones who have been retrained. Maria also believes it is more difficult for her to find a new job at her age. She feels it is unlikely she would be retrained on a new machine to improve her employment opportunities.

Cultural Historical Dimensions of Garment Workers' Learning

CHAT can help to make visible the ways that matters generally termed "contextual" and usually not spoken about directly are, in fact, constitutive of the forms of practice and learning in which people engage. For the garment sector, the key themes involve differential ethno-linguistic and gender effects on worker learning in interaction with the intensified class relations in an industry highly vulnerable to global capitalist competition.

Immigrants entering Canada are slotted into the Canadian vertical mosaic that differentially incorporates them into the dominant society. In her study of the Caribbean community, Henry (1994: 15) points out that there are "internal and external processes" involved in incorporating an ethnically and racially different group. The group's cultural patterns and familial and social relations, its internal processes, hamper the group's full integration into Canadian society. At the same time, external processes are at work "deny[ing] equal access to the goods and resources of society." Because certain ethnic groups are denied full incorporation into the dominant society, they become ethnically ghettoized. The Chinese community, for example, has developed its own business and employment networks in order to survive economically in the larger Canadian society. Therefore, Chinese garment workers are often employed by members of their own community who are small business people trying to survive in a cutthroat industry as well as in "the severe racial animosity of the larger society" (Bolaria and Li, 1988: 124). Often garment workers are reluctant to expose their employers precisely because they share a cultural bond of loyalty and they lack access to other forms of employment. The degree of occupational segregation by race is visible when all the members of a family are dependent on garment industry work as the only employment they can find without better English skills. For example, one of our respondents' brother is employed as a cutter and her three sisters are employed as sewing-machine operators. These conditions seriously limit general education and training provisions for immigrant garment workers.

From a CHAT perspective, language structures are the most important mediational dimensions of activity. Nowhere is this mediation more clear for adult learning than in the need for immigrants to learn a new language to operate in their new society. Immigrant workers with limited material resources often gravitate to, or are forced into dependency on, garment employers with little incentive to provide dominant language training.

223

Women of colour who were not born in Canada predominate in the garment workforce. They are among the most disadvantaged workers in the country, occupying the lowest paid jobs in a gender-segregated job market, while also suffering under a heavy patriarchal domestic division of labour. The further division of the female workforce into factory workers and home-workers is highly exploitative because of the vulnerable and subordinate position of minority women within the larger Canadian context. Employers take extreme advantage of ethno-linguistic minority women by deploying them as flexible workers and paying them below the minimum wage. These material conditions severely constrain the learning opportunities of most female garment workers.

We have already seen that the general activity structure of the garment industry has witnessed a massive upheaval affecting physical space (spatial dispersion of smaller factories and increasing home-working), intensifying patterns of inclusion and exclusion (extreme racial and gender-based occupational segregation and language discrimination), shaping social as well as material resources (older technologies, isolated and fragmented patterns of social interaction including rapid de-unionization), and intensifying production objectives (increasingly export-oriented). In terms of the objects of activity, workers are faced with many dilemmas that shape learning practice. For example, why learn one of the many specialized garment manufacturing skills if there are fewer and fewer jobs available, when those few jobs that are available often require multi-tasking generalists and when employers continue to rely on external skill development by new immigrants?

Immigrant women's factory sewing and home-working clearly represent among the most marginalized labour market segments which, in turn, shape the kinds of learning needs most relevant to the labour force, especially ESL and Adult Basic Education (ABE). These factors set the stage for the types of working-class communities that are associated with the garment sector (that is, largely immigrant ones with traditional patriarchal domestic divisions of labour, a heavy double day of labour for women workers and low levels of formal schooling). Training programs, as a key point of reference or "node" in workers' broader sets of activity systems, are far more likely to be oriented towards a basic survival tactic in the form of labour adjustment, for example, in the garment sector. In many cases, the union must fight directly for significant portions of its membership to even qualify for government-based educational resources. These activity structures seriously constrain the types of developmental trajectories garment workers' learning can take, leading to degenerative rather than expansive activity.

Regressive labour processes and managerial strategies, the kind we refer to as "adaptive" in reference to the work of Lazonick (1991), have a powerful and very specific effect on learning vis-à-vis the way activity systems function and develop. From the perspective of CHAT, the massive work intensification, speed-up, control and, in some cases, de-skilling in the garment industry turn the "object" of activity towards "survival" and away from "skill development" per se. Emphatically, this does not mean that skills and knowledge are not developed: garment workers are as active and intelligent in their practice as anyone – including corporate executives. Rather, it means that this development is integrated into production in the narrowest of terms and is oriented by some of the most basic matters of subsistence: e.g., time to spend with family, avoiding exhaustion while maintaining quotas. These are the skills that are, in a sense, the least likely to be openly recognized and represent some of the most hidden dimensions of the knowledge society.

Concluding Remarks

The Canadian garment industry has gone through a period of serious disruption and its future viability remains uncertain. Most workers in the industry face very uncertain job prospects. The barriers to participation in further education courses are extreme among garment workers and especially among the immigrant women workers who make up the majority of the labour force. Limited English-language skills and relative isolation present major barriers to some informal learning activities. Nevertheless, garment workers continue to be actively involved in a wide array of informal job-related learning activities which may be either taken for granted or more likely ignored by their employers, trade unions and even themselves. Hong, a recently arrived home-worker, expresses a common sentiment about learning among women immigrants when she says:

> I have to learn everything because I am in a new country. Many things I
> still don't understand about Canadians, their customs, feelings. I just let
> it go day by day. I sometimes go to the library to learn. But I have to sew,
> I have to make food for my family. So, I wait until they are asleep. Then
> I learn more English late at night. I don't want to sew for the rest of my
> life. It's a waste, sewing at home like in jail. I like to go out and meet
> people.

But as some sensitive local union leaders like Donna increasingly understand: "My co-workers do have a lot of knowledge and skill. They just can't speak English. They are very intelligent. They're not stupid, it's just that they cannot explain it in English."

Notes

1. The Ontario Ministry of Labour's definition of home-workers is: "Home-workers work in a private home for a business owned by someone else. They usually do such things as sewing, or other manufacturing; stuffing envelopes or other packaging; food preparation, assembly, repair or alterations." Employment Standards, Fact Sheet #14.

2. "As education providers, the central labour bodies play a very major role, particularly for those affiliates whose scale or priorities haven't allowed them to assign full-time staff to membership education" (Martin, 1995: 101).

3. This information is derived from the leaflet "Home-workers' Association."

4. Generalizations are difficult to make for the non-unionized workforce since it is fragmented and dispersed throughout Toronto.

5. The term "immigrant" technically describes the landed immigrant status of people immigrating to Canada and who have not yet become Canadian citizens (Ng, 1988). In the 1980s, common-sense usage of the term equated "immigrant" with "foreign-born" which "includes all persons who were born outside the country, regardless of current citizenship status" (Ng and Estable, 1987: 29). Today's everyday usage of the term increasingly refers to "ethno-racial minorities" which "hides a racialized concept of who a 'true Canadian' is and what 'real Canadian culture' resembles" (Das Gupta, 1996:55).

6. Canada changed its immigration system to a point system in 1967 that eliminated the previous discriminatory system that allowed people in racial grounds (see Bolaria and Li, 1988).

PART III

COMPARATIVE PERSPECTIVES
ACROSS CASE STUDIES

Household and Community-based Learning: Learning Cultures and Class Differences Beyond Paid Work

Introduction

> For men and women alike, the "home sphere" of household, family and
> community appears to be of central importance for their interactions with
> the labour market and with labour organizations. To understand the strat-
> egies of workers to improve their living conditions, we have to know
> what is going on in the "home sphere." (Kok, 2002: vii)

In one of our family interviews we met with Len Martin and Rose
Williams[1] at their home. Len, who is in his 40s, talked about how he
"taught younger workers the ropes" in the auto plant where he worked. He
taught them how to work the presses, change the dies, drive the tow-mo-
tor, keep their body parts intact and deal with the company and fellow
workers. This was all part of the job but, he explained that it had no relation to the company's training courses. Given the difficulty of talking on
the job, most learning happened after work in the parking lot, away from
the plant on lunch hours and quick trips on breaks to a bar named Caps.
Caps functioned as a classroom: none of the "curriculum" appeared in any
company course and all of the learning was informal.

From our research interviews we learned much from people, and much
of it didn't have anything to do with their formal work skills. We learned
about computer and computer repair, heart disease, religion, starting your
own business, politics, philosophy, labour history, Punjab history, Irish
history, German history, country music, and much, much more. In turn, at
points, we taught people things too: about sociology, playing the guitar,

marxism and even about the nature of academic research. During these interviews we became immersed in a collective, consultative, co-teaching, co-learning, informal learning network of ourselves, workers, their families and, in some cases, even their neighbours and friends. More than merely studying these learning relations we were engaged in them!

During our research for this book, we encountered a number of workplaces, union halls, homes and neighbourhoods. In every research site we met somebody like Len: willing to strike up conversation on any number of topics which they had somehow developed a working knowledge about. When one of us finally met his partner, Rose, we found out that she was the same – learning how to cut Len's hair better or trying to learn the "science" of baking the perfect loaf of bread, or how to navigate the bureaucratic vagaries of the state social benefits system – they both talked with excitement and expertise about their many different interests, some necessary, some simply pleasurable. Throughout, they told how they learned and taught others.

The "working-class" is obviously not monolithic. Our interviewees were not homogenous. Third-generation, Anglo-Saxon Canadian working-class households presented different challenges from those of new immigrants from Latin America or the Pacific Rim, young families grappled with different issues than did older couples, and each gender faced its own set of problems. Nevertheless it became very clear that, among other things, these diverse working-class households were particularly rich sites of learning. Most interviewees, unaware or unconcerned with the "research significance" of this finding, in effect, challenged the stereotypes of the working-class household as a bleak, stagnant educational context. Many people did cite a profound disconnection and resistance to schooling, but listening carefully and making space for the stories of working-class knowledge production revealed a great deal that has not been discussed in the literature to date.

This chapter deals with collective informal working-class learning in home and community settings examined across our research sites, with the intent of expanding our knowledge/power approach to learning practice. As we introduced at the start of the book, the learning described in this chapter seeks to satisfy our interest to explore the full range of learning practices (in paid work, in the union as well as in the community and home). It builds from the basic CHAT observation that particular spheres of activity cannot be thought of in isolation; that is, work-based, union-based, home- and community-based learning articulate with one another.

This chapter is broken into several general themes. We deal with family histories of working-class learning as well as the effects of the current context of (prolonged) economic restructuring. In general, the discussion below plays an important, general role in contextualizing the notion of a "dual economy" (Edwards, Reich and Gordon, 1975) adding depth to this concept in terms of its effect on learning practice. In terms of this dual economy concept, we can recognize "primary" ones made up of white, male, high-waged workers, often using leading edge technologies (e.g., auto assembly, chemical) and secondary ones involving those, often non-white, often female, workers who experience more insecure work at lower wage levels. For capitalism, particularly during intense periods of economic restructuring, we also recognize that these primary and secondary sectors of the economy, in a sense, are mutually constituting, inseparable despite unique characteristics of their own.

We should begin, however, by noting that there is a modest but well-developed historical literature on working-class family and community life. By the mid-nineteenth century, for example, with the manifestation of labour conflicts, European scholars and policy-makers began to observe the proletarian household. As van der Linden (1994; 2002) has noted, histories of working-class family and community life have rarely included discussions of how it relates to collective action through trade unions or other organized groups. However, virtually nothing within this literature has referenced the linkages between household and community life, and processes of learning per se. Despite this, writers like van der Linden (1994; 2002) have noted a number of dynamics that inform our discussion here. Paralleling the discussion in Tilly and Tilly (1981) van der Linden emphasizes the role of the working-class household as a key context through which "strategic repertoires" and "horizons" of possibility (or aspirations) form and operate. Through such observations we can begin to extend analysis towards the learning process (through use of the game metaphor):

> The 'players' have a limited number of pieces in their repertoire, and if these are insufficient a learning process is necessary before a new repertoire becomes available. In this way, the composition of the old repertoire influences the composition of the new repertoire. (van der Linden, 2002: 233)

Understanding the working-class activity, according to van der Linden, rests on labour relations and the institutional environments but, important for us, also the household and community networks. A type of working-class "network analysis" is referenced elsewhere as well (Joseph, 1983; Snow, Zurcher and Turner, 1962), though, again, excluding any specific "learning" perspective.

In addition, there is an array of commentaries on working-class families, although again only isolated fragments dealt with a broad conception of learning, with most of these focusing primarily on issues of social class in relation to formal schooling (e.g., Apple, 1990; Bourdieu and Passeron, 1977; Lareau, 1989).

It also became clear that the more informative literatures were those that set aside presumptions of a deficit in terms of working-class capacity, even if they did not focus on the learning process explicitly. These were detailed examinations (e.g., Rubin, 1976, 1994; Lareau, 1989; Hewitt, 1993; Luttrell, 1997; Seccombe and Livingstone, 1999; Luxton and Corman, 2000) that provided rich, qualitative and/or ethnographic data, giving a deep sense of the "endless process" of learning found in the lives of households like Len and Rose's above.

We try to use a narrative style here to reflect the complex interrelations within this research. The strength of these story-like accounts is that they help to integrate a wide array of factors and influences (material conditions, ideology/discursive relations, social histories) simultaneously present and operational in both the contexts in which working-class learning happens and the context in which it is researched and written about.

Home- and Community-based Learning Histories

Sarina Wilson works in an office in downtown Toronto and is married to Lionel, a chemical worker who had volunteered for the workplace-based learning portion of the project. She talked a good deal about her learning activities in general, but she also spoke directly and candidly of the differences between her and Lionel in terms of learning that she felt was directly connected to Lionel's working-class background. Sarina's own origins were (what she described as) "upper-middle class." Her father was a well-paid engineer and her mother worked full-time as an art instructor. In contrast, Lionel was brought up in a large African-American family on a factory-worker's wage. Lionel decided to leave school at age fifteen, dropping-out to pursue a career in music. A year later he was touring with a band, eventually gravitating to the more secure and better-paid cruise-ship and vacation circuit. This lasted on and off into his late twenties, but in between "tours with the band" he worked in a series of chemical industry jobs. Lionel met Sarina (from Canada) while he was working a vacation spot, returned to the Toronto area with her, married and, on the strength of his industrial work history, soon caught on at a local chemical factory – to start a "straight gig." To Lionel, life had always been "learn as you go" informally, first music and then chemicals. There was no need for structured courses.

232

Sarina: I guess I only realized this once I got married but I do have a certain class attitude ... I've noticed a slight difference between us. I think I'm different because I'm very self-motivated to learn and that's why I went to University. I believe in taking courses and bettering yourself and so on. I like to read a lot ... like my husband does not like to read novels or things like that. I've always been interested in culture, literature, art history. I really picked up a lot of that from university ...

Interviewer: Tell me more about how you might describe Lionel, the differences?

Sarina: Well just knowing how to, just administratively go through a process of finding the courses and what they cost and how to research them, I had experience doing this at university, and I've tried to encourage him to take courses.

Interviewer: He told me a little about his sideline business in multi-media stuff. He must do a lot of learning about that on his own?

Sarina: He's interesting and that's part of what I admire about him. He learns without courses. He learns, let's say picking up a camera or a computer and he will sit with it and figure it out himself whereas me I have to become professional on it first. I have to know everything about it before I even tackle it, so what is better I do not know. He has some different approaches though because we did some photography work together and he got me doing some photography work that I that I would not even have tried doing.

Interviewer: Without having read some or taken a course or ...

Sarina: I'd actually taken a course but I did not think I'd taken enough courses because I was not specialized and taught enough, but he *[Lionel]* taught me, I actually could do it. You just dive in without taking any courses ... I want to know every aspect and I get, you know, over-analytical ... Lionel and I come from very, very different backgrounds.... I actually credit him for giving me the courage to do this new stuff, just to dive in. I was applying to a Masters of Journalism when I met him and he kind of showed me, well maybe I can do it. Maybe I do not need the credential, after credential, after credential, and given what I like to do, and it's not rocket science. It's something you just learn to do and you can only learn them by doing them. You can't learn how to be on camera by reading about it. You can't. You have to be on camera.... I want to be doing things versus just absorbing information.

Sarina summarizes what many interviewees had expressed in their comments on the relationship between class histories and learning. She outlines class difference in learning. Her upbringing and learning is described as "analytical"; "self-motivated," aimed at "bettering yourself," and having interest in "culture, literature, and art." These make up her basic sense of class difference in terms of learning.

A key point we want to introduce using Sarina's comments relates to the relationship between discourse and ideologies of learning on the one hand, and material practice on the other. In the lives of the working people we interviewed, learning usually did not revolve around a subject area that was neatly "packaged up" and abstracted away from the world: interviewees insistently declared to the degree they could claim the term "learning" as their own, they learned in the "real world."

The identification of these ideological and discursive dimensions of learning in our interviews is significant. Antonio Gramsci (1971) writes about the significance of the distinction between thought and action, and draws broader implications.

> The contrast between thought and action, i.e., the co-existence of two conceptions of the world, one affirmed in words and the other displayed in effective action, is not simply a product of self-deception. Self-deception can be an adequate explanation for a few individuals taken separately, or even for groups of a certain size, but it is not adequate when the contrast occurs in the life of great masses. In these cases the contrast between thought and action cannot but be the *expression of profounder contrasts of a social historical order*. It signifies that the social group in question may indeed have its own conception of the world, even if only embryonic; a conception which manifests itself in action, but occasionally and in flashes – when, that is, the group is acting as an organic totality. But this same group has, for reasons of submission and intellectual subordination, adopted a conception which is not its own but is borrowed from another group. (327)

What is important for us here, in Gramsci's description is the contradictory existence of thought, language and action within hegemonic relations, and this can be applied to the way the people in this research think, talk and act in regard to their learning. It speaks more or less directly about the relationship between the learning that working-class people do as separate from the learning (which is more easily identified as "learning" per se) done by those in dominant class groups or in settings identified as legitimate sites of learning. As we describe throughout this chapter, knowledge, power and difference become intertwined through everyday language and practice with notions of social class and learning.

Following feminist and Marxist-feminist writing we could apply a discussion of the relations of "naming" the world that writers including Hartsock (1987) and hooks (1984) have helped to develop. The theoretical concept of naming one's experience can be closely linked to basic understandings of power relationships represented within (and partially reproduced through) language as discourse. As Spender described it, naming is "the means whereby we attempt to order and structure the chaos and flux of existence which would otherwise be an undifferentiated mass. By assigning names, we impose a pattern of meaning which allows us to manipulate the world" (1987: 163). As hooks (1984) has written, naming one's own experience (but not naming it in isolation) is an essential part of liberation when done in conjunction with political action; together forming the basis for new knowledge forms, and theory grounded in the material positions of oppressed people. And those familiar with the work of Freire will find these ideas quite familiar in that the "generative themes" that were the building blocks of a pedagogy of the oppressed essentially revolve around the same principles.

But "naming" is not a simple matter. We are not free to name as we choose and we are always working within, or at least starting from within, the confines of the history of discourse and ideology as well as specific material circumstance.

Sarina's cross-class observations are not the only entry point to analysis of the connections between social class and learning across home, community and life histories. Other people interviewed also discussed these connections and their current learning practices. An autoworker, John talked about a vivid memory of community-based learning and how, growing up in a small town, a local teacher affected his understanding of "learning" just before he dropped out of high school.

> So we went up to his house and helped him move furniture and he wanted a wall ripped out and there was like five of us – all his students – and we were going to rip out a wall and then frame it, and stick it all back up. He says, 'Okay, this is how you do it,' and drew it all out and showed us ... and we had it up and drywalled. You know? Like this type of stuff. After that I thought, "Gee, what do you do when you do this? What do you do when you do that?" Then, when I wanted to figure something out, I went to friends. Because one guy, he knew all about furnaces because his father was a furnace repairman. Another guy's father was an electrician. All right? So you draw on the knowledge of someone else. (John)

He would draw on this notion of learning throughout his life.

Enzo, the self-described son of hard-working immigrant parents who came to Canada with little education when he was just young, is now a millwright in a chemical factory. He describes the joy of learning in a neighbourhood friend's driveway while growing up in Greater Toronto.

> When I was ten or twelve years old, one of my best friends when I was growing up, his father was a mechanic. We'd learn a lot about cars, go-

carts out in the driveway just fooling around. You learn a lot, just trying to build a go-cart with a good engine, and stuff like that.

Informal experiences like these were consistently cited as very positive and important and help ground an understanding of class-differentiated understandings of the learning process. Working-class learning in the examples above is distinctly collective and, significantly, thus also counters dominant/bourgeois ideologies of individualism. In this way, community-based learning is not only an important part of people's present practice, it is also part of their life histories as well. Subordinate groups, in absence of access to legitimatized notions of knowledge and learning, looked back and searched out the experiences in our interviews that could define a positive expression of both learning and their own standpoint in the process.

At home, Jane Brooks (a chemical worker) reflected on her main learning interest and how she came to take it up.

Actually my family was all musical. My oldest brother – he played lead guitar, rhythm guitar, the same. My little brother used to play drums. My youngest brother got into it when he grew up and played lead guitar. My sister plays piano and organ. My dad used to play the accordion. So I do not know. When I came up here I just enjoyed the music …

Tony, a garment worker and union representative has a variety of other informal learning activities that rival his time and energy commitment to his paid job:

I've been involved in coaching sports including soccer, bowling, tennis, outdoor sports like hunting. I do furniture carpentry which I learned on my own, desks, wardrobes. I did not take a course. I use my own imagination to learn. I watch and work with my friends or whatever. People say I make beautiful things. I also love dancing. Dancing is my favourite hobby. I'm always learning to be better. Always trying to learn more. I cook my own meals every day and I've learned some different ways. I'm also reading books to help my daughter with her math studies. My interest has always been to learn.

And, Greg, an autoworker fond of betting on the horses, explains:

I do it myself. [The computer] just keeps track of the data. I've got two databases, four word processor files, two or three spreadsheets, and a couple of other things I use. I've been on this method now for four years … I'm a good winner. I've got all the books. I study. I just don't walk in and bet. I study all the angles you might see. It's not that complicated, you know, it's a matter of finding a halfway decent way to rate the horses and observing what's going on. I'm mathematically inclined. I like math.

236

Math can tell you everything. Everything you want to know, math will tell you. Horseracing is a science. It's all mathematical probabilities.

In general terms, people spoke of many positive connections between their learning at home, in the neighbourhoods and wider communities while growing up. The quotes above are representative of experiences that people felt helped them to define themselves as learners now. Seldom if ever did people speak about schooling in the same way. More often school was used as an example of how not to learn, with people citing, in particular, its tendency to individualize people in the context of competitions.

The Connections Between Economic Restructuring, Home and Community Life

Pete Donaldson was a key informant for the research at the chemical factory. He had been very involved with the union local and had met with us virtually on an "on-call" basis to confirm bits and pieces of information. We heard about the plant's worklife, health and safety issues, training procedures and company history. In addition to telephone conversations, we had met several times for extensive interviews at a restaurant close to the plant. Pete's life had been fairly rough –he dropped out of high school at a young age, abandoned an attempt at a trades certificate, failed at his own business, suffered a substance abuse problem, and experienced the sudden death of a family member – but he felt that, within the last decade, things had settled down considerably. He had settled into life at the chemical factory, married his partner, Susan, and in their mid-30s they had purchased their first house. We were invited to their new home in an unfinished subdivision for an interview. One thing we had a chance to talk about was the need for Susan to work outside the home. As most of the people had said in this research, in this period of economic restructuring it now takes two incomes to maintain a household, and the statistics discussed in earlier chapters clearly bear them out. The types of economic changes identified in the statistics reveal themselves in particular ways in working-class learning, community and home life. Pete and Susan, for example, have decided not to have children at this point, and, as we see below, Susan's work in a large bank has become a good example of what the statistics reveal is happening in most workplaces – demands on time and energy of workers is being ratcheted up.

> *Susan*: Well I've been there for seventeen years. Started as a teller and I've done a lot of different jobs. I've kind of gone up the ladder a bit, but now I decided to go back down *[laughs]*. I find it better at the bottom... Yeah, I do ... well I was a sort of the branch manager's assistant. I did

that for about eight to ten months and didn't like it. It was a lot of work and not very much pay and a lot of aggravation.

Pete: And a lot of hours.

Susan: Yeah, and a lot of hours. I didn't enjoy it. I've never stayed with the bank because of the salary because the salary they pay at the bank is very poor ... So I ended up taking a different position to get out of the hassles, so now I just sort of go in do my job and leave at the end of the day ... It allows me some freedom on the weekends, I don't have to bring work home with me or stay late at night, and there's something to be said for that.

They go on to outline further intensifications that are actually forcing Susan to look for a new job in which the combination of work/learning intensity is more reasonable, or at least partly paid for.

Susan: I'm actually trying to get into Ford to work on the line ... Yeah! The pay is a good portion of it, but no more of the bullshit being expected to put in 110 percent everyday. Being expected to do all these courses on my own time and all the other politically bullshit stuff that goes on at the bank, you know?

Pete: Like the stuff for the United Way, it seems like you always have to go in on a Saturday and wave banners and bake cakes. There always seems to be a bake sale going on and all this crap. Like she doesn't want to do it.

Susan: Well, it's not that I don't want to do it, it's that I don't want to be made to feel that if I don't do it that I'm going to shoved down that corporate ladder even further and the fact of the matter is that you know it is expected of you that you give your 110 percent. They even quote that to us you know, "100 percent is not good enough. We expect 110 percent from you" and if you're not willing to do it, not willing to produce it then we'll find someone else who will. It's just really that cut and dry.

Like other interviewees Susan feels the weight of restructuring not so much on wages which have always been "very poor," but in the terms of work intensification. There is always pressure, Susan says, to do more. Most importantly in terms of class-based learning though, Susan describes how she feels a good deal of pressure to continuously take courses and do readings at night or on her lunch hours for which she is not paid. This is the somewhat hidden world of classed, informal learning exploitation that is generally unrecognised. These employer demands also extend to Susan's sense of "voluntarism" as corporate charity fund-raising programs and events are shifted primarily onto the backs of workers, from organizing funding drives to baking cakes and selling cookies. The super-exploitation effect from Susan's point of view is intense enough for her to be actively seeking to leave. To improve her quality of life she is actively looking for a place on an assembly line where she feels union representation would help reduce this stress.

The Distribution of Free Time and Learning in the Home

Many of our interviews gave an important glimpse into the way home and learning lives are intertwined. Most people identified a real "time crunch" in their lives. It shapes learning in the home and community generally but also hits the female partner the hardest in the form of the "double day" (Cockburn, 1988; Luke and Gore, 1992; Rubin, 1994) – a full day of paid work plus primary responsibilities for housework. One study estimated that male partners have as much as 12.25 hours of discretionary time per week *more* than their female partners do (Communications, Energy and Paperworkers, 1994), and another estimate puts this difference at twelve hours per week (National Research Council of the United States, 1991: 43). Still another indicated a huge seventeen hours per week difference (Mara-Drita in Rubin, 1994:256). Statistics like these have implications for any number of issues – among them the ability of people to have some control over their free time and to participate in learning activities of their choice.

Lillian Rubin has been involved in the study of working-class families for over twenty years, first with the book *Worlds of Pain* (1976) and then with her follow-up *Families on the Fault Line* (1994). She comments on the shift that she has noticed in attitudes around housework and gender:

> In fact, it may be that over the last two decades, the greatest shifts in relations between men and women have taken place in working-class families rather than in middle-class ones. Two decades ago I found few working-class men who would even give lip service to the notion of gender equality, whether inside or outside the house. Today many of the men I interviewed are quite sensitive to the needs and wishes of their wives. It's true that this sensitivity often is not translated into action. But the very assertion of the ideology of equality by men who resisted it so thoroughly before is itself a step forward, the first perhaps in the struggle for genuine change in the family. (Rubin, 1994: 78)

Rubin did not discuss the issues of learning specifically, but using some of her analysis as a springboard, there are a number of key factors perhaps the most basic of which is the structuring of time that must be considered: how is time divided up in the working-class home and community; who gets the time to do what they want when they want; and who gets the time to learn? The last portion of the previous section draws attention to some of the research on the division of free time between men and women that revolves around the complicated negotiations of household chores.

Rosa Cantella, working as a line worker in the small parts factory, examined in Chapter 6, was interviewed with her husband Frank. Rosa and Frank lived in a duplex with their three boys and Rosa's parents, not far from where Rosa works, about an hour's commute to Frank's job where he works as an auto mechanic. They have been married for many years and met in Canada, having immigrated separately from Italy when they were each in their early twenties. Rosa knows the rigours of home and child care well. Although her mother's help is a great relief (she does at least one load of laundry every day), for Rosa the double day is still a challenge. As a teenager in Italy Rosa's parents were forced to leave the country in search of work. Rosa was left to manage the household and her younger brothers and sisters on the factory-worker wages her parents were sending back to her from Germany. Separation was very painful for the family and they eventually reunited. Rosa then began the wait for her twenty-first birthday when she could legally leave for Canada without her parents' permission. A quarter of a century later, Rosa and Frank are happily married in Canada and living together with their extended family under one roof.

During the family interview at their home much was discussed about learning and finding (free) time around the home. Frank and Rosa had each come from what they called a "traditional upbringing." They explained that by this they meant that the man was the "breadwinner," and the woman was the "homemaker." With the changing times, however, they both discussed how people have had to start looking at things in a new way; but this is not to say that they have been able to leave the past behind. The experience is different for every couple but it is clear that various generational effects have a good deal to do with the negotiation process around housework and the idea of traditional upbringing. Here Rosa and Frank talk a bit about their views on the matter.

> *Interviewer*: Should men do more housework?
> *Rosa*: [*responding quickly*] If the woman works and the man doesn't or has part-time, I don't see anything wrong with the man doing housework.
> *Frank*: [*disagreeing*]
> *Interviewer*: [*to the husband*] Why?
> *Frank*: The woman is better than the man in the house.
> *Rosa*: I know guys that are very good in the house.
> *Frank*: I help my wife. I have no problem with that.

Rosa: You both have to try to do the best you can ... I know some people where the woman works full-time, husband works part-time, and they help each other. It works out.

Frank: *[agreeing]* It works out if you want to stay together.

Rosa: Everything is a part of education and learning what people have, like marriage. Couples today, you see marriage end after a few years because no communication.

Frank: It's not that.

Rosa: To stay together – one has to be a little bit stupid; one has to be a little bit smart. I tell Joey *[her son]* that one day when he gets married, I don't want him to sit in front of the TV while the wife is in the kitchen. That would work 100 years ago. Today you have to stick together.

Although Frank agrees with the general principal that women and men should try to "work things out" and stay together, he hesitates to give up the housework split that sees men come out on top. He just wants people to stop fighting. He prefers to view the "working out" process as dependent on a couple's general commitment to staying together. Rosa's perspective and experience, on the other hand, makes her much more aware of the problems that exist in the division of housework. For Rosa staying together involves developing fairness in the sharing of responsibilities and ability to negotiate or communicate about this split. In the quote above, she makes an interesting slip which was not unusual in our interviews. Despite her understanding of the need for a dual income and some sort of renegotiation of the division of household work, the couple in her example only begin to "help each other" as a result of the husband's part-time (versus the wife's *full-time*) job.

Other couples in the research spoke equally candidly about the split in housework and its effect on free time, confirming much of the existing research literature on the issue (Bacca Zin and Gitzen, 1993; National Center for Research, 1991; Rubin, 1994). There is certainly a shift in people's sensibilities around who should be doing what, but as Dan Harris, a chemical worker, is quick to point out, people's talk may still not translate directly into actual changes in practice.

In general I think it's mostly women who do it ... But it's not like that here. Most of the guys that I know, whether their wife works full-or part-time, the woman still does a lot of the work ... I think it goes back to the way you were raised. (Dan)

Two other male chemical workers, Enzo and Lionel, provide descriptions that support Dan's thinking. "We try to split but the wife does it more than me. I grew up and my mom did it all ... I just don't think of doing housework. I don't notice things to do ..." (Enzo); and "I do some, but it is the wife's domain because that's just the way it works out. She cares more about it than me" (Lionel).

The organization of household labour and caring work has important implications for discretionary activities of any kind and this includes planned and unplanned learning. Ron, a chemical worker, outlines the arrangements that his family has developed in response to economic context.

> Well I can say that, for me, my wife and I have changed roles recently. I've become a house-father. I'm actually the mom of the house now and the wife has a job that takes her out of the house anywhere between ten and twelve hours of the day. She's also now taking a course at the college. As a matter of fact she's there right now. I've allowed that to happen because I can look after those things which she used to have to do because of the balance in the house. Not everybody can get away with that, I understand that sometimes it's the thinking, because the way we're brought up, some men today say that women have their place. Slowly, we're learning that we have to share in the raising of the children, but in our house I've learned a lot and we are getting to that point where now she can go off and do what she wants to do. I've pushed her to do that because I knew she needed the challenge for things to be more fulfilling. So I opened the door for her, so now more doors open up because of it. So it's an attitude in the house that creates that.... [in speaking of a co-worker who has no children and who is doing a university degree part-time] We're similar in one sense. He's doing what I'm trying to do, that is, continue my education part-time only I'm on a tighter schedule. So where Jimmy may be getting "quote unquote" sufficient sleep, say six or seven hours a night, I have to operate on three ... [after his night shift] I got to bed at noon today and I had to get up by 2:30 [for the interview] so I've had two and half hours of sleep. But that's what I guess I have to do for now to do the things I want to do, and do the housework and go to work.

Another interviewee from chemical factory explains that,

> Well it's just the nature of the family, the make-up of the family, the age of your children if you've got children especially with two people working, to fit that into your day. You know I know some guys at work who get to bed every night at nine o'clock and you know they get upset if they end up getting to bed at 9:30 because they're really tired the next day. Shit, I haven't been to bed before one o'clock for, well what my daughter's six now. You know you've just got so much to do, you don't have that free time. Either of you. Especially the female in the relationship just because of the upbringing. (Tom)

From all of this some important general observations can be made. Most importantly, shifts in the economy clearly affect material conditions and social arrangements. The most basic example here is the need for the dual income. As almost everybody said it is a central factor in the shifting ways in which people arrange their lives. Although Rosa only alludes to the new social and economic context "*Today* you have to stick together" it is clear that she and Frank are in the process of a renegotiation of the "traditional" arrangements with which they both grew up. Other, younger couples, are responding to the cultural/material shift by changing more dramatically while still other couples continue to maintain the gendered division of labour which produce a woman's double day and a gendered imbalance in free time for things like learning projects. A second part to this negotiation can also be clearly seen in the comments of Enzo and Lionel; and, perhaps, even in the more progressive talk of Pete and Susan. Being a breadwinner or a homemaker is also an identity to which people are deeply attached and which helps them understand and act confidently in their social world. Despite a conscious sense of the inequality, trying to alter these identities can clash with images, skills and an awareness that have been developed over many years. Lillian Rubin comments on these effects on the working-class families she studied.

> Why is it so different now? One answer surely lies in the realm of ideology. The ideal of the good mother that lives so powerfully inside us is relatively new one, an ideal type that reached its zenith in the aftermath of World War II, when the economic needs of the nation came together with the psychological needs of a war-torn population. Then, the economy needed women to leave the work force and head the drive toward a greatly expanded consumer society. And people, starved for the family life they had put on hold during the war, hungry for the goods they couldn't buy before, were happy to oblige. But the old image of mom in the kitchen with an apron tied around her middle no longer fits either the economic or psychological lives of families today. Universally, the women I interviewed work because they must. Almost as often they find a level of self-fulfilment and satisfaction on the job that they're loath to give up, even in the face of guilt and anxiety they suffer. (1994: 81)

Clearly the realm of ideology has effects in the material world through household practices. It becomes very obvious that the social practice of free time and housework is a type of negotiation that both men and women (albeit not equally) play a role in reproducing. Frank actually sums up this resistance, collusion and reproduction nicely when he tells us that, "Men have to learn that things have changed and women have to learn to let go."

"I can't think of anything – can you?": The Invisibility of Working-Class Learning in the Home

Finding our way off the highway and into one of the many different working-class neighbourhoods of Hamilton, Ontario we pulled past rows of small war-time housing that proliferated in Canada in the 1940s. This is the new neighbourhood of Dan and Debbie Harris, one of the many outgoing, generous and energetic couples that have offered so much to this research with working-class families. As we get out of the car, they both appear from the backyard gate and greet us on the driveway; Jamie, their daughter, in her mother's arms and Buddy the dog follows close behind. All of us make our way inside and settle into the living room for the interview with the television on in the background.

Dan and Debbie are both in their early 30's and originally from the east coast of Canada. Dan has worked in the chemical industry on and off with different companies for almost ten years and has also completed a community college diploma. Debbie is currently working full-time as a receptionist in a local health centre. Dan and Debbie rely on a combination of Dan's shift work (night shift lets him take care of Jamie during the day), Debbie's aunt, and Dan's grandmother each of whom live in town. This lets them care for Jamie while maintaining the double income they need for their mortgage payments.

This ability to juggle their lives is demonstrated in the interview when they cope with their young daughter and their dog while still managing to provide a calm and coherent description of their lives as learners. However, like almost all the other interviewees in the course of this research, the first time we ask what they have learned in the past year the response is same.

> *Debbie*: I can't think of anything. *[turning to Dan]* Can you?
> *Dan*: We really haven't had too much time. I'm either working or with her *[Jamie]*. I'm always on the go basically.
> *Debbie*: I take her *[Jamie]* to the park. Go for walks. Colouring. We watch Barney. I have to keep busy with her, or she gets mad … but you know I can say we learn a lot about babies and basic child psychology from dealing with Jamie.

When we first begin to discuss learning, people often deny that they are doing any learning at all, or alternatively they are not very sure what you mean by learning, or what really "counts" as learning. Susan and Pete, introduced earlier, put it this way:

> *Susan*: We've had to learn a lot, but I don't know if I'd say I learned about it?

Pete: Well, I guess I wouldn't say we've studied it, but ...
Susan: Like I've read up on it because we had to come up with a design, and we had to do some investigating as to what would be good for our area but ...

Sometimes people asked plainly, "Well what do *you* say is learning?" As practical and straightforward as statements like this might seem, they highlight a very important point: "there is no such thing as "learning" *sui generic*" (Lave, 1993: 5–6). Interviewees frequently denied that they were in fact learning. To engage people with the concept of learning as a broad participatory activity, we would often give personal examples such as learning to fix the car, attending an afternoon health and safety session or learning about computers with a friend. People would generally then begin to recognize that we were not just talking about some form of abstract knowledge in the library somewhere, but instead the kind of knowledge that made up the fabric of *their* lives as mother or father, as auto workers, bowling team members, soccer coach or dance-organizer. Despite how people usually talk about it, learning is inseparable both from who is doing the learning and, as people like Jodi point out, who is judging that learning. Thus learning has a differentiating and documentary dimension. It must be surfaced from the level of the tacit to be registered and it is in this process that an important process of differentiation occurs. As in the case of Susan and Peter above the questions become: whose reading is considered "studying;" whose learning is privileged, whose knowledge is considered worthwhile? This dynamic of denial/realization appeared frequently and, as alluded to in the Gramsci quote, it is a pattern that is not mere coincidence but is linked to deeper, social historical struggles. In Dan and Debbie's case, after some initial confusion about the nature of learning, they begin to describe the home-based learning they do together.

> *Debbie*: I guess I've learned a lot about the car now that we've been running into problems. I've learned about what the oil does and what the temperature and the radiator has to do with it *[laughing]*.
> *Dan*: She's had some breakdowns.
> *Debbie*: But Dan has showed me a lot and I listen and watch what the mechanic does.

And, with the new house, which required a good deal of renovation work, though Debbie again denies any involvement initially, there are clues as to the depth of knowledge and skill development going on:

> *Interviewer*: So did you both do a lot of renovation stuff around your house or what?
> *Debbie*: No, he did it all.

Interviewer: So you just stood around or ...?

Debbie: Well, I helped out like with the plumbing. I helped laying out how the pipes would go and stuff like that, lining up the stuff. I learned a lot there.

Interviewer: That must have been a good thing for your relationship eh? *[everybody laughs]*

Dan: Well with plumbing you have to make sure that ...

Debbie: You really have to be right on, and I learned where all the shut-offs *[for the water]* are ...

Interviewer: Well that's good to know.

Debbie: And I learned where the electrical box is and I know all the stuff on that now. And I know how to do the furnace now.

Dan: It broke down.

Debbie: And the repair guy came in and I watched how to oil it and fix the air filter and clean it with the vacuum...

This partner to partner learning is central to the home- and community-based learning profiles that were generated from the WCLS research, and home maintenance and renovations in particular were a very popular activity in which working people learned and developed skills together while at the same time saving a little money.

Other prevalent topic areas were also learned collectively in the home including health issues centred on ill health in the family or extended family. Chemical worker, Sean, and partner Donna, for example, have much of their living and learning structured around Donna having been diagnosed with a serious disease of the nervous system in the late 1980s.

Donna: It is mild at this point but it is progressive ... I guess I have a bad time of it every two or three years, so we just make arrangement for these times. I take time off work, once I had to take off almost a year.

Sean: We think that she's developing a pattern of about every four to five years but it is also stress related....Of course we had to do a lot of learning all the time about the disease and how we can make little adjustments.... But really we just live a normal life like everybody else.

Pamphlets, books, discussions with friends, other patients, doctors, nurses, watching television shows, etc. all make up the curriculum that Sean and Donna have been mastering together for almost a decade. This curriculum involves practical information that they use directly, but also includes more abstract theoretical information including how the disease develops and acts on the body in light of new observations on genetic and environmental factors. Sean and Donna start to dovetail this knowledge with Sean's decade-long informal studies of, and engagement in, workplace health and safety activism through his union that, in turn, extend to Sean and Donna's interest and concern for environmental issues. This "dovetailing" of learning reflects the way that different social spheres of activity overlap and inform one another (e.g., work, union, home). As the interview with Sean and Donna proceeds the focus turns to recreation and they begin what turns into a lengthy discussion of their two winter vacation spots, Miami and Las Vegas. While they talk about the information finding, sorting and processing they do around their trips, we find that it is a learning experience for us: "How do they get the dogs to run around the track?" "What does it mean to double-down?" Sean and Donna's eyes light up as they explain their (elaborate) theories of betting at the track and casinos.

Working-Class Community Living and Learning

Home- and family-based learning practices were not the only ongoing, deliberate learning activity that people described. Beyond learning with immediate and extended family members, people often talked about how they learned among friends, neighbours, team-mates, or members of various clubs and organizations. These learning relationships can be seen as different from those in the workplace and home because of the amount of choice or discretion that people can often exercise in them. While people could not generally move easily in and out of neighbourhoods they did have a good deal of choice over who they were friendly with and with whom they took part in learning activities. In this way, the term "community" is more than a topographical or physical distinction of a neighbourhood boundary. It is a group with which people chose to be involved on an ongoing basis, in this sense drawing on a sense of community defined, more or less, by members' discretion. Community has, over the years, taken on a number of different meanings.

The people in our research were involved in many different types of communities including the traditional neighbourhood. To return to Len and Rose whom we met at the start of the chapter, often the learning that went on between neighbours could be as simple as getting together periodically under the hood of the truck to figure out a problem together.

> I'm pretty good mechanically, but the only thing that's starting to confuse me a little bit is these new cars with the computer in them ... I've been having a little trouble with this one [his truck] but between me and the guy next door we have been, he's sort of a mechanic too and it's not too bad, but we've been working at it together. Like, it's almost impossible at the same time though because unless you got a computer to hook up to that you can not really do too much. The only other thing is that I learned how to re-program one. (Len)

Similarly, Sean Gilbert describes getting together with a neighbour periodically such that he now says he is becoming "handy" around the car.

> I recently learned how to replace my muffler myself. I learned from a neighbour who is very mechanical. I learned how to change my radiator with the same guy actually. We kind of learned together, between the two of us the job got done. I have a pretty old car but I can do a lot of stuff myself now.

These seemed like friendly collective moments of experience and learning that took place within an ongoing development of, in these cases mechanical, knowledge. At the same time, however, many of the same themes discussed earlier, such as the dominant way of talking and thinking about "learning" in connection with working-class people, are still evident. As we posed the question earlier – what is the relationship between so-called abstract and non-instrumental learning and the working-class; and in this context, what are the relationships that help structure thought and action? What, for instance, is Sean's primary goal? Is he developing a broad knowledge base of mechanics or is he more concerned with the fact that the "job got done," or does it really make sense to separate out these dimension of the activity at all?

As we mentioned earlier, Pete and Susan Donaldson had recently moved into a new neighbourhood complete with the odd vacant lot. What we did not see driving up to the house was a large tract of land behind the house which they told us was being turned into a "park."

> *Susan*: We were very involved in the building of a park that was supposed to be going in behind us. We all started a neighbourhood committee and that's been going on for a while now. To make a long story short, we decided it would be a good idea to get our hands into the development of the park.
> *Interviewer*: How many people?
> *Susan*: Initially it was about twenty people, all neighbours, but there ended up being about four of us who are really keeping up ... It really helped us get to know other people in the neighbourhood.
> *Interviewer*: You must have had to do some reading and looking into bylaws and stuff around this park issue?
> *Pete*: Actually we had a bylaw initiated.
> *Interviewer*: So you must have had to do some learning around that?
> *Susan*: I guess so, like we had to attend council meetings and we didn't know how any of that worked. But I wouldn't say we went out to the library and did any training or anything, it was just sort of a go and get our hands on really.

As Pete and Susan went on to describe, as a collective neighbourhood group they had talked to lawyers, attended municipal council meetings, organized committee meetings for information sharing – but all this, according to Pete and Susan, was still somehow not exactly "learning."

Jane Brooks helps describe other types of discretionary communities with which working people are involved. Earlier in the chapter she described the large and musical family in which she grew up. Not surprisingly, she is now involved in what could be called a local community of musicians. While this is generally centred around the different bands she is playing with at different times, it also extends to include any musician with whom she comes in contact. These relationships, whether old or new, almost universally revolve around what each can learn from the other. Jane begins by describing how she became involved with her current musician friends, and the relations of learning that are involved.

> We used to go down to the pub every Saturday night – there used to be a whole group of us you know – and then this guy that used to have his own band he used to get people up to sing. And we used two or three songs and you know then after doing that for a couple of years he said "Well come on, we'll get you up every set to do three or four songs. So I kept on doing that for a long time. And then I started taking music lessons – like playing guitar right? ... And then there's learning all the different songs. I mean I have to sit down here and learn the songs. And then if they're not easy songs you can't expect the other ones to *pick it up*, so I mean if you want to sing it they have to sort of know it basically, so you hopefully give them the tape and they take time to learn it.... [in the band] Everybody was wanting to learn. We worked good together, associated good together. It was just great ... I never gave anybody a hard time. They always used to say – you're too easy going! [laughs] "Get on our case more. *Make us learn more!*" (Jane)

Marcel Longpré, a French-Canadian auto worker, is the same way. He plays guitar in his church band and speaks fondly of any opportunity to learn from other musicians. Sean and Donna Gilbert recounted how the bowling community is a social club but at the same time one that requires a lot of learning.

> Our main hobby is bowling and there's a lot of learning in this activity ... It's a little bit like the "team system" at work ... I'm a "team player." Really the whole sport is about constant adjustment. We bowl about three or four nights a week. (Sean)

They read different technical magazines and books, but they also learn from people who "really know," just watching and "picking their brains."

As several people made clear, learning is not the only reason to be involved in different types of community and collective interaction, and there were very few "communities" in which learning per se was the central purpose. For example, beyond companionship, most of those in the research obtained their current jobs based on involvement in these communities. Roberto heard of his current job at the parts factory from friends in the local soccer club, rooted in a cultural community of new immigrants from Columbia. His partner Maria brings in extra money through her part-time work in a toy factory, which she learned of through a neighbour in her apartment block. Jane got the lead for her current job based on connections with a community of former co-worker/friends with whom she still spends time – not to mention that she made contact with her current band at the bar she and her partner used to frequent.

Family-Based Union Learning

Talking over coffee in downtown Toronto during our interview with Sarina Wilson, we learned how uninformed she felt about her husband Lionel's activities and involvement with the local union. Lionel had been a union steward in his department for years and Sarina had never really heard much about this aspect of his work. But over the last couple of years, an old work injury was starting to bother him again. The local union helped facilitate a number of changes at work including "light duty" but things seemed to have been getting worse despite this. As we discussed earlier, in a period of economic change that requires a double income for a decent standard of living, people are understandably nervous about anything that jeopardizes their situation. While Sarina was deeply worried for Lionel's health, it was in this context that her interest in learning more about the union began to grow.

> *Sarina*: Well, mostly now because he's injured ... before he was injured there would just be these union meetings and his work hours would somehow be directed through the union but I really didn't know much at all. Now there's the whole issue of compensation, and what job could he have and will they keep him and will the union support the claim. It's difficult because I don't understand it. I don't have an affiliation to it really. I have no control. It's really hard.
> *Interviewer*: Now would you be interested in any sort of seminar or presentation about the union or what they can do?
> *Sarina*: Yes, I really would. I would like to know what they're about so I can understand what he's doing.

Here learning by whatever means possible, like learning about computers or learning the dominant language, while interesting, also takes on an economic imperative.

Several couples provided colourful examples how informal learning about the union comes into the working-class household. Dan and Debbie Harris discuss the role of a member of the local union executive as being important in this respect.

> *Interviewer*: What about politics? Do you guys learn anything about politics?
> *Dan*: Well, I work with Tom *[a member of the local union executive]* so I learn something everyday *[laughing]*.
> *Interviewer*: What about you Debbie? Does he bring back information to you about this or anything like that?
> *Debbie*: Yeah, all the time.
> *Interviewer*: What kind of stuff?
> *Debbie*: He tells me everything … I learn a lot about unions … I guess I learn how critical it is for a person be in a union and I guess what some of the advantages are …
> *Dan*: Well, I don't want to go into a bit like Tom does here but I think people don't understand unions and what their forerunners have had to sacrifice and everything. And people don't understand how things are being slowly taken away from them so subtly that they don't notice …
> *Debbie*: Like an example would be that some friends I know that aren't unionized are afraid that if they become unionized they will lose their seniority and that they'll have to start from scratch.
> Interviewer: So they just don't have the information then?
> *Debbie*: That's right…. I actually learn a lot from just being around *[with Dan]*.

Around a kitchen table Pete and Susan shared how they teach each other about unionism. Pete tries to teach Susan how to resist her bosses at the bank and Susan teaches Pete how difficult resistance is in a non-unionized workplace like the bank.

> *Susan*: Like I'll come home and try to explain it to Peter and being that he's in the union it's kind of hard for him to relate or understand.
> *Pete*: The crap that she puts up with there.
> *Susan*: Yeah, like the stuff that's fed to you and you just have to take.
> *Interviewer*: What's Pete's relationship in helping you identify the bullshit?
> *Pete*: *[all laugh]* Well, sometimes I just shake my head and walk away!
> *Interviewer*: It seems like you learn something from his experiences at his work though.

Susan: Learn something? I learn a lot about the strength of unions and as compared to working at the bank. There's a lot of little things that, I'm trying to think of a good example ...
Interviewer: How much do you know about Pete's union work?
Pete: More than she wants to!
Susan: ... he likes to talk about his work. If there are things that are important with him that he thinks should be shared with me then he'll tell me about it ... Just about everything actually.... He goes on courses and I hear about that. He's involved in grievance and I hear about those, sometimes not always. If he's spending a lot of time on it usually I'll hear about it.... To be honest until *[to Pete]* you were part of a union I didn't know what they were about and I had a negative thing in my mind because people in my family never had anything good to say about them. I don't think people know anything about unions unless their family is a part of them.

The learning process between Pete and Susan did not stop at just thinking about the issues. Instead, the discussion merges with concrete action: agitation in the workplace.

Pete: I think four years ago or something and I saw in the paper that the bank made something like a billion profit and I cut it out of the paper and highlighted some things, and I said to her "Go on post it up at work," "They'll fire me!" she said.
Susan: I posted it up, but I didn't say who did it.
Pete: I'd be marching around with it showing it to people.

Learning about unionism in the home was not reserved only for adults; children often learned about work and the labour movement as well. Some of the children with whom we had a chance to talk seemed to have the types of aspirations that one might expect as typical of children. However, the life of the most active union members clearly spilled over into the life of the children. During one interview session with Tom, an active member of a local chemical workers' union executive, a view of this process came into focus. Interviews at Tom's home turned into an evening full of food, drink, lively radical politics and unionist folk music. Tom's brother, a unionized worker himself and an activist with a local anarchist group, sat clapping Tom's baby boy's little hands together to the beat of the guitar. Near the end of the evening I asked Tom's eight-year-old daughter what she would like to be when she grows up. Perhaps sensing the mood, she responds without missing a beat, "a union leader."

Concluding Remarks

> It's only experience that teaches you, and your experience is written in
> bigger letters than what's been told you by the school teacher. Aye, it's
> in bigger letters. It stands out clear like this [and he made a gesture with
> his hands]. You see what it means, but you did not see at the time. It's
> only afterwards. (Retired factory worker, from Jackson and Marsden,
> 1986: 76)

Using the research data we have talked a fair amount about family histories, economic contexts and specific home and community-based learning activities that go on in and around working-class households. We have
also discussed a good deal about ideological contexts and the ways people
talk and think about learning, i.e., discursive and ideological relations. We
have tried to do this by speaking from a working-class standpoint.
Amongst all this, however, we have yet to make any definitive statements
as to the character of working-class learning.

The opening quote for this section deals with class difference and it
represents a thread that runs through virtually every reflective description
of learning that people provided in this research. The old man in the quote
above sketches out a sense of difference that involves three major components: "learning"; "schooling" and being "working-class." Our point is
that, as well as looking at historical material conditions, much can be
gained by investigating the most shared senses of differences that working-class people themselves express. Along with most of the people in this
research, the major point of difference that the old man in the opening
quote expresses involves the meaning of "experience" as distinct from
"learning" that happens during schooling.

As Rose and Len pointed out earlier, a lot of "learning" that working-
class people do is not recognized as "learning" – sometimes not even by
working-class people themselves, and, when they do use the word "learning," there are visible efforts to "steal it away," as it were, from the abstracted, classroom-bound usages. This act of appropriation, of "situating"
or "particularizing" learning, has been a central element of a working-
class definition of learning for some time and it makes clear that abstracted knowledge on its own simply will not do.

> Like I like to cook, you have to know that, okay, I'm just applying it here,
> okay taking a recipe for bread, like bread is a science so they say, cook-
> ing is creativity but bread is a science. Okay I take a recipe from a friend
> of mine and I tried it, I did, I tried to reproduce it. But one day I thought,
> "To heck with this, I'll come and watch you." Well, I wrote down every
> bloody thing she did, every step she took, everything. Then she said,

"This does not seem quite right it needs a little more of this and a little more of that," and then she says, "It has to be like this and it has to sound like this." I mean, this stuff was not in this recipe, excuse me [laughs]! And I thought, I got four pages of writing with little tips and hints and this sort of thing all the way through, but it's exactly what he's [Len] talking about. *You just have to be there* [laughs]. (Sara)

One of the first extracts we examined in depth came from Sarina Wilson. Having grown up in an "upper-middle-class" setting Sarina described all sorts of classed "differences" much as Pierre Bourdieu perhaps would have, as a series of absences.[2] Almost in one breath Sarina tells us working-class learning is not particularly "self-motivated," aimed at "bettering yourself," "analytical," or interested in "culture, literature, and art." Susan and Pete Donaldson indicate several times throughout their interviews that although they have "learned a lot" they often were not sure that they have really "learned about" or "studied" anything. And in discussions of learning in the home-based learning section, Jodi, a parts worker, hits the nail on the head by asking "What do *you* say is learning?" All these are strange and curious statements to any person thinking that learning is learning, but as noted earlier these are not individualized cases of "self-deception" or confusion. Whereas within dominant class culture, "learning" exists in relation to a huge array of words, theories, buildings, fond memories, certificates, regalia, jobs and promotions – these working-class people do not, or at least in this research have not, drawn on this dominant reservoir of meanings. Instead they seem much more comfortable describing the importance of learning in their lives as "experience," "getting the job done," "just sort of go and get your hands on," or sitting with one's child "just shaking our heads." These are indications of specific class relations that are revealed at the level of discourse: a dominant discourse in which the standpoint of subordinate groups is denied, denigrated or ignored.

Certainly the working-class people we spoke with have a sense of the class differences within conceptions of learning. Among the most organized and secure working-class groups – generally middle-aged, employed, semi-skilled, white, male union members – in the primary sector of the economy – they are not helpless before the dominant conceptions of learning, despite the fact that interviewees clearly know there is a definition of learning and intelligence "out there" that does not include them. They know that to some people and even at times to themselves they are unlearned or uneducated or even stupid.

Len: It's just like I'm not sure what they are going to do in the year 2000 when the computers do not reset their dates properly and run out of numbers ... Computers are going to crash in the year 2000 because they are going to run out of numbers.
Rose: Look at me. I do not even know how to turn one on! ... *[to Len]* How do you know so much?
Len: *[pretending to whisper to Rose but then speaking up so I can hear]* I'm telling you I'm suppose to be stupid! *[Rose laughing loudly]*
Interviewer: They keep telling you! *[laughing]*
Len: I know, I know, I have to listen! *[all laughing loudly]*

We must, of course, resist the temptation to romanticize these hidden dimensions of the knowledge society. We note that it has been difficult to fully assess the full complement of work and learning relations among workers in the peripheral sectors of the "dual economy," such as the garment industry, or those with dispersed family lives living within unstable and highly fragmented communities. In addition, for other interviewees, while a set of individual interviews with them about their work-based learning was acceptable, there was anxiety and resistance to having interviewers enter too deeply into their personal lives. For these and other reasons, the hidden dimensions of working-class learning as it involves community and home life remain hidden to a degree.

Nevertheless, from this exploratory examination several new observations emerge and many key points raised in other chapters are re-emphasized. Historically, broad community-based learning and organizational networks have been associated with some of the most progressive periods of working-class mobilization in North America and Europe. Generally speaking, the most effective periods of the working-class mobilization in the union movement in Canada have been associated with expansive views of education. The colourful memoirs of Allen Schroeder (1984), for example, provide rich, grounded descriptions of an activist's life in which we see a rich overlay of different spheres of learning and practice. Recalling activity in the 1930s and 1940s, Schroeder talks about how he and others in the city of St. Catharines (Ontario) formed informal study circles to support both union organizing as well as a Co-operative Commonwealth Federation (Canada's socialist party of the period) cell structure. Like the CAW union now, Schroeder, then a member of the UAW, strongly endorsed and spoke at length about the importance of linking informal learning across the home, community, political party and union hall. Freisen (1994) offers an informative comparison of educational philosophies of Schroeder and others in Canada during this period, widely recognized as a "high-water mark" in terms of activism and progressive mobilization. In this comparison there is the suggestion that narrow conceptions of learning and its role in the labour movement in fact coincided with the emergent trend of stagnation and bureaucratisation of the labour movement later in the post-World War II era. Today, the CAW's Family Education Centre is a living symbol of an expansive view of learning that once was more dominant on the labour scene and which may even be again. It is accounts like these that are perhaps the most useful starting point for sensitizing ourselves to the practices of working-class learning in the home and community.

However, times have clearly changed since the post-war "boom" decades in North America. The intense time pressures – in the home, for example, where the dual income is a necessity –and the increasing average length of the work day for Canadians (Lowe, 2000) sometimes make discretionary learning difficult. Work has become more insecure and for those left in downsized plants it has become intensified. These take their toll on the mind and body and, in turn, affect the ability of people to come alive to learning in the home and community.

Making strategic use of the expansive array of working-class learning capacities across home, community, union hall and workplace, however, requires at least a minimal ability to recognize the learning as valuable. Moreover, it requires an analysis with the ability to identify where it still flourishes, where it does not and why. In this chapter, we have seen that the recognition of learning capacities, by analysts and practitioners alike, can be difficult. Unlike the earlier period that Schroeder recounts, there are powerful and broadly transmitted ideologies of "learning," "life-long learning," and "knowledge work" not to mention an expanded dominance of schooling in the lives of ordinary people. Each of these things seems to contribute towards the hierarchical ordering of learning practices in such a way as to ratify deficit valuations of working-class capacities.

Notes

1. "Len" and "Rose," pseudonyms, were interviewees from a factory studied but not included in the research site chapter in the book.
2. i.e. "Working-class children have no access to [so-called] cultural works, so the teaching of culture always anticipates an experience of culture which is missing." (Bourdieu, 1994: 9)

CHAPTER 9

Surfacing the Hidden Dimensions of the Knowledge Society: The Struggle for Knowledge Across Differences

> Knowledge is power, and knowledge in the hands of working men and women is power that will change the world. — Inscription on the plaque of "Labour's Home," Wortley Hall, South Yorkshire, U.K.

Dimensions of the knowledge society that begin from the lives of workers are obscure to most employers, most researchers, many unions, and far too many working people themselves. If we are to "surface" these dimensions for progressive, democratic social use in the interests of working people, then we must begin, as we have attempted to do in this book, with the fullness of their learning lives. The inscription at Wortley Hall may not be in the style of the majority of academic researchers, it may be beyond the purview of pragmatic policy planners, and it is clearly antithetical to the latest business gurus but it is worthwhile reflecting on for our purposes. Learning is about change and it is from a working-class standpoint that the most serious and fundamental questions of change come into view. Our research has tried to take a close look at learning for paid work and beyond paid work from this standpoint. We chose five diverse sites for case studies. The studies have shown many differences and similarities in terms of learning practice across key economic sectors in Canada. Lessons from these studies will, we believe, be useful to progressive researchers, policymakers, working people and their organizations in other countries as well. However, there remain several questions that can only be answered through comparative analysis. Perhaps most important of all is the matter of recommendations and practical steps forward.

The comparative nature of this chapter provides the opportunity for further exploration of our knowledge/power thesis on learning. Across the five site chapters we have argued that power and context are constitutive features of the learning process, rather than separate from it. We see that power, as an ideological, cultural and material relationship, comes in myriad forms. Power relations can be experienced as a form of domination or a form of resistance and countervalence, an issue that revolves around the concept of standpoint. Our standpoint is that of workers and their or-ganizations. Roughly speaking, where union strength is at its greatest we see developed forms of countervailing power relations and the most com-prehensive *lived* opportunities and support for workers' learning in all its forms. Therefore, for example, the auto site with its powerful and well-developed union traditions is associated with expansive lived opportuni-ties and support for formal training and continuing education as well as informal job-related learning. The notion of "lived" opportunities and sup-port is vital to our claims in this area as it helps clarify the difference be-tween workers' actual learning and employer representations of "learning organizations."

The CHAT approach encourages attention to social contextual differ-ences as integral to the learning process. Our knowledge/power thesis rec-ognizes the mediational importance of several aspects of power in shap-ing workers' learning practices. These included industrial sector market strength and managerial strategies, union strength and the critical role of ethno-linguistic status, gender and age in shaping learning practice. More site-specific aspects of knowledge/power relations are also relevant. In the college site, for example, we outlined how, since the workplace was also an educational institution, it offered high levels of opportunity and support for workers' continuing education participation. Likewise, in the chemi-cal plant, we noted progressive rhetoric on learning and skill development among workers, enhanced partially because of work dangers in the plant and the need to pay particularly careful attention to health and safety. At the same time, in both of these sites actual *lived* access and support for learning was highly segmented with white-collar administrative workers (college) and those in the "high performance" work groups (chemical) having generally privileged access. Finally, largely due to the market tur-bulence in their sectors, the garment and small parts manufacturing sites featured the greatest struggles of all. With union strength being weakened in these sites, the knowledge/power thesis would suggest that these work-ers are particularly vulnerable, with the least lived opportunities to engage in discretionary learning and apply their knowledge and skill. But we have also found that even in the most oppressive employment conditions, workers somehow often find the time and interest to engage in a substantial array of learning activities.

We will attempt to pursue these points to put more comparative flesh on our knowledge/power thesis, beginning with a discussion of patterns of use of formal schooling, participation in organized training and further education courses, and engagement in informal learning across the sites. We follow with a discussion of gender, ethno-linguistic status and age effects on learning practices of the organized working-class and conclude with a brief discussion of recommendations from the site research.

Uses of Formal Schooling, Organized Training and Informal Learning Across the Sites

Formal Schooling Patterns

The college site stands out as the most receptive in its recognition of formal schooling. The employer, being a credential-granting institution, seemed to make recognition and use of formal credentials slightly easier in practical terms. However, even here there remain serious inequities in relation to educational certification of different occupational groups. Administrative officers, coordinators and technical staff often engaged in formal courses at the college and outside, including university, professional and even graduate level degrees, while the maintenance, custodial and, in particular, largely female cleaning staff were not actively encouraged to do so. The latter's approach to formal schooling, their formal attainments and experiences, and their perspectives on avenues to further education are not significantly different from those who share similar challenges at the other workplaces, particularly those in the small parts manufacturing, garment factories and home-based garment piece work.

The attainment levels and workers' approaches to and experiences within formal schooling were similar across the auto assembly and chemical worker sites. These stable, high-wage sectors with relatively desirable jobs seemed to be particularly prone to credential inflation, however. In the auto assembly and chemical sites, it was not unusual to see trained accountants, engineers, teachers and a variety of other post-secondary and community college graduates walking the floor side-by side with high school graduates and drop-outs, and even elementary school drop-outs from an earlier labour market era. A variety of factors account for this credential underemployment, especially the failure to recognize foreign credentials. In addition, the relatively less secure, low-waged environments involved comparatively lower formal credentials, constituting the peripheral segment of a dual labour market (Edwards, Reich and Gordon, 1975).

The distinction between primary labour markets composed of good jobs, relatively good pay and decent working conditions, and secondary labour markets composed of poorer quality jobs, poorer conditions, and poorer pay is intimately linked to the relative power of employers and workers. The more power employers have, the less value workers' formal schooling will likely attain (see Livingstone, 1999a).

Still focusing on formal learning, we also see that trades credentials were treated in very different ways from formal educational credentials. The personal belief in the value of trades credentials and apprenticeship learning was strong among those workers who possessed them. Indeed, the (mostly) men would often speak of (formally) long-completed apprenticeships as still ongoing: "I'll always be learning" said a trades worker whom we interviewed in the small parts manufacturing plant. Millwrights, tool-and-die makers, machinists and associated maintenance positions (electrician, plumber, carpenter, pipefitter, etc.) felt that their credentials could not be separated from their work, nor, in many cases, from who they were as people. Clearly, one factor that made a significant contribution to people's assessment of the value of these credentials and, by extension, ongoing trades learning, was the large degree of freedom to organize their own work, to take up specific learning tasks either by themselves or, more often, in conjunction with other trades workers. Closely related was the substantial bargaining power (on the general labour market as well as informally and formally in the workplace) that trades workers exercised through their credentials and knowledge, in addition to their chronic undersupply. In fact, the example of trades credentials offers a fascinating comparative case in relation to the perceived usefulness, meaning and exchange value of non-trades–based formal credentials. For most of the other workers in this research there was a tension between the different meanings of formal learning and credentials. This difference was drawn out most clearly in the chemical and auto assembly chapters in which workers outlined how learning activity in school has often been of only minor, intrinsic value in itself.

Most interviewees indicated that formal schooling was of variable value. More specifically, its value comes in four general forms: 1) as a credential, 2) as a type of self-esteem, 3) as basic level skills, and 4) as a strategic element to support informal learning in collective systems of activity/learning. First, regardless of their immediate, practical value, credentials are useful as a commodity on a labour market (either internal or external to the workplace). Second, formal credentials are valued by workers as providing esteem in relation to dominant ideological and cultural meanings. The small parts manufacturing, chemical and household-learning chapters dealt with this theme in a fairly explicit manner, but its principles are operative across all sites. Although some workers seemed to effectively reject dominant meanings of formal credentials (e.g., in the chemical site), more often workers talked about the personal sense of "pride" associated with obtaining a diploma or degree; or, alternatively, the personal "shame" of incompletion. We can see that formal schooling was an institution form through which significant cultural meanings that express and help reproduce class-based power relations were at play in the lives of workers within and beyond the workplace. As Len, the auto parts worker, commented in the final section of Chapter 8, he was "supposed to be stupid" largely because he did not complete his high school diploma. The third form of value ascribed to formal schooling was rooted firmly in basic skills like reading, writing and math, in terms of daily skills required to do a job. Basic literacy and numeracy skills allowed workers to do their work in production, in the union, in their home or community, more effectively. Finally, interviewees thought formal schooling could also have value in relation to forms of knowledge that could not be accessed through informal learning networks, self-study or trial and error. For example, learning about electronics in the course of maintaining one's machine (in the auto or small parts manufacturing plant) or batch-mixing equipment (in the chemical plant) is important, but sometimes it helped to speak to someone who was formally trained in this area. Another key example of this complementary value is when computer learners engage in informal learning collectively, perhaps talking with a worker who has some formal computer training in the area of software design. In each of these examples, formal training had value as an additional resource. Moreover, *school-based learning* was used as a means to facilitate and strengthen existing *informal networks* (the latter being more fundamental to the broader current learning process).

Organized Training and Continuing Education Patterns

Previous surveys of adult education have found that unionized workers are more likely to participate in formal courses than non-unionized workers. They are also more likely to have the costs of these courses supported by their employers (e.g., Sawchuk, 2003b). When workshops of short duration are also included, these patterns are sustained for both industrial workers and service workers (e.g., Livingstone, 2002). However, as the site chapters should suggest, there are also substantial variations amongst unionized work sites in the extent of provision of formal education and training programs.

Perhaps the best way to compare the character and comprehensiveness of training and continuing education offerings at our different sites is to view them under the general categories of what are called tool courses (e.g., health and safety, stewards), production process courses (e.g., high performance production, team-based production, overviews of production process), general skills (e.g., ESL, reading/numeracy, computers), and, social issue courses (e.g., anti-racism, anti-sexism, political economy). An overview of the training and continuing education landscape at each site is provided in Table 9.1. The table outlines interviewees' shared perceptions of the availability of courses in each category[1] as well as the sponsorship (company, union, joint) of the course (i.e., who ran the course). It is suggestive of the actual differences across different sectors and firms. Some additional courses may be offered but if a representative cross-section of the workers is unaware of the courses, they might as well not exist. In the case of the chemical workers, for example, we learned from one of the steering committee reports that the company offers a computer lab and instruction for workers as well as salaried employees. Indeed, the regional office of the union also offered a computer course. However, none of the workers we interviewed knew of these offerings. The general awareness of courses may be suggestive of the degree to which the development of a learning culture in the workplace is part of the collective consciousness of the workforce.

Table 9.1: Training and Continuing Education Across Research Sites

	Tool	Production	General Skills	Social Issues
Auto	U, M, J	U, M	U, M, J	U, M
Chemical	U, M, J	U, M	U, M	U
College	U, M	M	M, J	—
Small Parts	U	—	J	—
Garment	U	U	U	—

Legend: U = Union courses said to be available; M = Management courses said to be available; — = no course said to be available; J = joint programs said to be available.

Of course, the aggregate profiles summarized in this and the following tables are based on very small numbers and can only begin to take on relevant meaning when considered in conjunction with the in-depth interview materials. Nevertheless, we do see some important effects of the strong union and workers' learning culture at the auto assembly and chemical site. This type of learning culture effect combined with various gender, race and age effects (favouring white, middle-aged males) operates across sites to help account for the particular pattern of distribution that we see in Table 9.1. A strong union culture helped workers appropriate notions of learning that included them despite class positioning and limited formal schooling. Positive effects of the relatively secure employment conditions in these two sectors also play a role, resulting in a general heightening of worker consciousness and control around the issue of learning.

As Table 9.1 shows, in terms of breadth of training and continuing education, the auto assembly site, influenced by the strong educational culture of CAW seemed to provide the greatest programmatic opportunities. The difference between sites is indicative of a range of interrelated factors including the relative wealth of the company and union, the culture of the company and the union. A highly developed union culture has consistently put training and education near the forefront of CAW planning and negotiations, resulting in possibly the most developed educational, training and learning *union* culture in Canada (Martin, 1995). One way that we can begin to confirm a judgement of the development of a union's learning culture is in the descriptions of its workers around programs and issues of workers' learning. Interviewees consistently cited a variety of company-sponsored, union-sponsored and jointly-sponsored programs. In the case of union-sponsored training and continuing education, workers could typically point to courses at multiple levels of their union (national and local) as well as courses offered in connection with centralized labour bodies such as the provincial and national federations.

Chemical workers appeared to have significant company-sponsored formal training opportunities in both "hard" (e.g., statistical process control, or health and safety courses) and "soft" (e.g., communications, personal relations or team courses) skills. However, we should be careful to distinguish between the existence of training opportunities on paper and real-world training and continuing education. The chemical company in our research actively bills itself as a high-performance, training-oriented company, but it is clear from talking to workers that this is, at least partly, an exaggerated claim, with a preponderance of OJT replacing more formal training and continuing education courses per se.

The college site was quite strong in its training and continuing education offerings and although these opportunities have not lived up to their progressive potential (due to radical restructuring), the college's offerings remain comparatively high. Also affecting the general level of training and continuing education among the college workers is what appears to be a unique formalized learning culture element. Relatively convenient access to courses and daily exposure to an educational environment may encourage wider acceptance among workers of organized educational programs.

Perceptions of availability of courses at the small parts and garment sites are generally lower than at the other three sites, probably reflecting both generally lower course provisions and more difficult economic conditions to support training programs. The fact that forms of (temporary) labour adjustment programs are in place in these cases muddies the water somewhat. Levels of formal training may appear higher under these conditions than they normally are and, among workers, the programs are seen less as developmental opportunities than as "parting gifts" to laid-off employees. Partly due to their involvement in this research, the small parts manufacturing workers with support from their union were actively engaged in creating a joint educational committee structure and piloting specific programs, including computer skills and ESL, with an eye towards bargaining for a innovative, *worker-led* "production process education" course. Union-based training and continuing education was, de facto, the only game in town for garment workers whose employers (or contractors in the case of home-workers) provided few training opportunities themselves. These union-based programs revolved around a fairly centralized structure (both UNITE and the Home-workers' Association) with special emphasis on basic education in the factories and among the growing numbers of home-workers.

Basic education programs were probably the single most salient training and continuing education issue for workers across sites. While it is much less of a concern among the (Anglo) auto assembly, chemical and most highly credentialled college workers (administrative, clerical and technical staff), ESL was seen otherwise as a key formal and informal learning issue. Custodial and housekeeping workers at the college, and a range of both garment and small parts manufacturing workers in particular saw learning the dominant language (often adding to an existing multi-language competency repertoire) as key to their work and union activity, as well as their home and community lives. ESL instruction was seen as a gateway to fuller participation. The comparative lack of attention it received amongst employers should be taken as an important signal of the general gulf between interests of the company and the interests of workers.

Figure 9.1 summarizes actual course participation in terms of the percentage of workers who have taken at least one job-training or other continuing education course or workshop in the year prior to their interviews. The overall participation rate is similar to the rates found for unionized

workers in the most recent national survey of course and workshop participation (Livingstone, 2002). But the data confirm that it is the chemical and
college organizations, as declared learning organizations, and the auto assembly plant, with the strongest union culture, that have the highest rates
of formal participation. The basic participation rates are somewhat higher
in the chemical and college sites where nearly all workers have been invited to some sort of orienting workshop around recent restructuring initiatives by management. Only a minority of small parts factory workers
took a job-related course and an even smaller proportion of the garment
workers did so.

The pattern for non-job–related courses is somewhat different. The
auto workers' extensive education provisions encourage the highest participation rates, followed by the college workers. Chemical workers have
markedly lower rates of non-job–related courses than their job training
rates. Small parts workers again have low rates, in this case virtually no
discernible involvement in other courses. Garment workers have rates almost as high as auto workers; however, virtually all of this participation is
in English language and other basic skills courses, usually subsidized by
government programs.

Figure 9.1: Course-taking Across Sites (% answering yes)

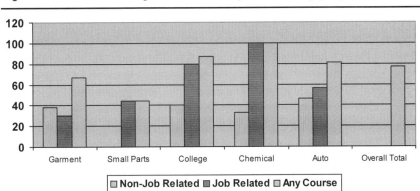

*Note that the response for Non-Job Related Courses under Small Parts Manufacturing
is zero.

Table 9.2 offers better estimates of the actual extent of involvement in
further education courses at each site. It summarizes the average number
of hours that workers spent per week engaged in training and continuing
education, both home-based and job-based. As in Table 9.1, we see that
material resources, dual-labour market and union culture effects intersect
to produce a consistent pattern of distribution. Auto workers appear to

have the greatest opportunities and take the greatest advantage of these opportunities in comparison to other sets of workers. Once again, the weakest unions in small parts and garment have the lowest general amounts of time. But is also notable that the chemical and college sites, with high levels of participation, have much lower average hours than auto workers, suggesting that course provisions within these "learning organizations" may be more superficial or restricted.

Table 9.2: Course Taking Time Across Sites (average hours per week)

Workplace	Home-Based Courses	Job-Based Courses	N
Auto Assembly	6.7	4.5	14
Chemical	0.4	1.5	15
College	1.5	2.5	15
Small Parts Man.	0	0.5	9
Garment	0.5	0.6	22
Weighted Avg.*	1.8	2.0	75

* Averages weighted based on total number of codable interviews in sample

Informal Learning Patterns

The very limited previous comparative research on workers' informal learning has found that workers in general are just as likely as other occupational classes to devote time to job-related and other informal learning activities, the most recent self-reported estimates generally being around five hours per week devoted to job-related informal learning (Livingstone, 2002). These preliminary estimates also suggest that non-unionized workers generally are no less likely than unionized workers to spend time in job-related informal learning. A simple explanation for this pattern is that all adult learners have relatively large autonomy to determine if and when they engage informal learning activities, in contrast to the greater resource and prerequisite constraints involved in formal course participation. But a comparison of our case studies through the CHAT approach to learning perspective suggests that there are likely to be systematic variations in the informal learning patterns of organized workers according to some of the same contextual differences effecting participation on formal courses.

As we outlined previously, learning is rarely assessed throughout its full range of variation, and even more rarely is this range assessed from the standpoint of workers themselves. Filling these gaps means, amongst other things, looking more closely at the cracks and crevices where workers are more able to exercise discretionary control. In our study, workers have been seen to engage in a enormous range and amount of individual and collective informal learning activity. But these learning activities remain constrained by the most extreme demands against their *time, space* and *energy*. These three basic material constraints form the foundational structure within and against which working-class people continue to struggle for and engage in learning activity. This struggle shapes the basic patterns and culture of learning among workers. Nowhere were these structural constraints more pronounced than in the lives of the garment workers. Here the effects of lack of time, money, energy, and even the ability to associate with other workers, were considerable. Yet still virtually all working people persisted in their learning plans and efforts. For example, garment workers, as well as small parts manufacturing workers and the cleaning workers in the college, described how they were learning English from the television at night, borrowing English tapes from the library, even listening closely to English speakers in their daily lives. Especially in the absence of courses, this informal learning was in many ways essential for social and material survival in Canada and in this period of restructuring.

Drawing on a CHAT approach to learning, we can outline the degree to which working-class job and home life is grounded in informal learning activity. People commonly band together with neighbours and friends to engage in issues ranging from community development to cultural festivals, car maintenance, forms of worker resistance, and beyond. Family members band together to learn, cope and act on family health problems, for example. Co-workers band together to teach and learn about computers, health and safety, and workplace issues through their union local. People take up learning projects that are deeply influenced by and in turn influence their everyday material and cultural lives. These findings confirm the most basic claims of the CHAT approach: that learning is disproportionately determined by cultural material practice and not strictly determined by forms of individual cognition; that it occurs wherever and whenever people have the opportunity for genuine participation in a system of relations, or activity system; and that these systems are shaped by mediating artefacts, history and the specific social standpoints through which people engage in practice.

Prior research has suggested, and evidence in this book confirms, that informal learning and tacit knowledges on the job account for the largest share of production-centred learning in any workplace (Betcherman *et al.*, 1997; Kusterer, 1978; Livingstone, 1999, 2001; Sawchuk, 2003a). Informal learning can be constrained by restrictive organizational structures (e.g., surveillance, neo-Taylorism, lack of opportunities for input and discretion) as we saw with the small parts manufacturing workers, or through the isolation effects on garment home-workers, but these structures do not make informal learning disappear. Rather they funnel it elsewhere: into forms of resisting responses as described by auto worker Ben Hamper (1991), or productively into forms of union activism, community or home life. But this rejection of workers' human capacity to contribute always takes its toll. Writers since Adam Smith have noted that continuous production lines are the most blatant means of achieving this effect. In our site chapters we saw, for example, how production operators in small parts manufacturing were actually agitating for more knowledge of production when the opportunity to interact and discuss with other workers was restricted. In this case, management experienced acutely the contradiction between wanting more knowledgeable and engaged workers while at the same time wanting to retain hierarchical structures of control. The consequence was that workers, through lower wages, intensified work and regular lay-offs, were largely drained of their interest towards job commitment.

The relevance of managerial strategies on worker learning was evident throughout. Lazonick's (1991) claim that more adaptive and innovative regimes differentially shape relations of knowledge and career trajectories within firms was generally confirmed. A basic relationship between patterns of informal learning and firm type can be gleaned by extending this analysis. Thus informal production learning is either the object of suppression under adaptive and cut-throat strategies of firms, or informal learning forms an important "coat-tail" on which innovative, worker-knowledge integration strategies of firms can succeed. Across our research sites, with the exception of the chemical site and more selectively in the college site, most organizations seem to be engaging in forms of adaptive and cut-throat approaches. This is clearly the case in small parts manufacturing with downsizing, multi-tasking and intensification. Similar adaptive management strategies are apparent in the garment industry, with the addition of attempts to revive widespread sweatshop conditions. In auto assembly, lean production methods take a similar basic approach, despite the fact that it is resisted and otherwise mediated by a strong union learning culture.

Table 9.3 provides a comparison across sites in terms of estimated time devoted to intentional informal learning per week. Once again, sectoral effects, union culture effects and a host of other matters appear to shape the amount of (home- and job-based) informal learning that is reported. While self-reports of informal learning cannot, in themselves, reflect the full range of informal learning (much of which is tacit in nature), it becomes quite clear that matters of sector stability, actual free time, personal valuing of one's activity as a legitimate knowledge/skill production process, and strong unions play important roles in how workers perceive informal learning.

Table 9.3: Informal Learning* Compared Across Sites
(average hours per week)

Workplace	Informal Home-Based Learning	Informal Job-Based Learning	N
Auto Assembly	11.1	10.4	14
Chemical	3.5	3.4	15
College	4.0	4.1	15
Small Parts Man.	4.6	5.1	9
Garment	1.7	2.3	22
Weighted Avg.**	4.6	4.7	

* Responses begin from a 1hr/wk minimum; reports capped at 20 hrs/wk.
** Averages weighted based on total number of interviews in sample

In terms of job-based informal learning, auto workers tend to spend much greater amounts of time learning than do workers at the other sites. Much of this informal learning revolves around labour rights issues and is not necessarily centred on technical job tasks. Their strong union culture and negotiated protections facilitate high levels of discretion in learning about general employment issues, despite the strict constraints that assembly-line production imposes on their job performance. Interestingly, the small parts manufacturing workers, many of whom also work on an assembly line score quite high as well. As the site chapter indicates, this is partly due to the extensive involvement in technical change-based learning by male trades and maintenance workers with considerable discretionary control. However, also related is the fact that this is a multilingual workplace where most workers speak with pride and enthusiasm about gaining functional literacy in two or more different languages in the course of production, despite resistance from management and the organization of the labour process. Close behind are the college workers. In this case, it is the relative autonomy of a range of occupational groups that encourages these levels of informal learning. Chemical workers have somewhat lower averages, again in contrast to their universal participation in workshops. It seems as though workers in such learning organizations are not necessarily more motivated to engage in job-related learning than are workers in other organizations, in spite of the appeals of management. Further down the list in terms of job-based informal learning are the garment workers. Here an array of forces align to make discretionary and open participation in systems of activity difficult. These include extreme working conditions, turbulence and instability in work, the destruction of secure work lives, a weakly organized, or in the case of home-workers unorganised, labour force, not to mention the double days of these predominantly female workers.

The patterns for home-based informal learning are quite similar. Once again, auto workers lead the way. The chemical, college and small parts sites all share similar averages. Garment workers indicate the lowest averages. The long arm of the job appears to extend to influence home-based learning, probably through the time, space and energy effects produced by the job site and its associated learning culture.

The general pattern of intensity of participation in both courses and informal learning across these sites is most directly explicable in terms of relative union strength. The auto plant workers have enjoyed the fruits of a militant and highly effective union that has negotiated relatively good wage for their socially homogenous membership. Overtime remains a drain on free time but they have the resources to do what they want. In addition, their strong union-based learning culture seems to have had an important effect on their assertion of the value of their own forms of working-class learning practice. These workers are more likely to identify what they learn and the ways they go about learning *as* a legitimate and valued form of knowledge acquisition. At the other extreme, the sectoral fragmentation and associated weakening of unions in the garment industry have undermined the heroic efforts of the merged union to engender a shared learning culture in a highly diversified workforce. Various other factors mediate worker-learning patterns at the other sites but they are generally intermediate between these extremes in terms of union strength.

Gender, Race and Age Effects on Workers' Learning

Workers' learning activities are also influenced by gender, race and age (see also Chapter 2). Below we deal with each in turn.

Gendered Dimensions of Working-Class Learning

The site chapters outline quite intense patriarchal structures of learning. As Chapter 8 illustrates, declining real wages and the notion of the "family wage"/"secondary earner" affect all members of the household but particularly women. Men can take control of much of their space, time and discretionary energy for learning activity as their women partners take primary responsibility for preparation of meals, laundry, care for children, etc.. However, the chapters also begin to point to several key local mechanisms upon which larger patriarchal structures depend, against which resistances may be focused, and which are largely formative of gendered working-class learning practice. The most prominent, empirically substantiated items in this research are the female double day and sexist workplace promotional practices.

A variety of scholars have noted the functioning of the female double day (Gannage, 1986; Luxton, 1980; Rubin, 1974; Sargent, 1981). It refers to the female partner's primary responsibility for household tasks despite the increasing numbers of women who also have one, and even two, paid jobs, often in low-wage job ghettos (Livingstone and Mangan, 1993). In our research, this basic patriarchal mechanism is most clearly viewed in the chapters dealing with a relatively high female:male ratio, i.e., in small parts manufacturing and the college and the garment sector in general. The most striking examples were the female small parts line workers who, exhausted and often sore from their efforts at their outdated workstations, make their way home to care for children and to prepare dinner for their families. Other examples detailed the simultaneity of garment homeworking (which includes picking up and delivering materials out of town, machine maintenance, and rough bookkeeping), home-making and child care. Faced with this heavy workload, it is little wonder that women workers say that discretionary learning is sometimes more of a luxury than a necessity.

For many employed working-class women, learning activity becomes shoehorned into whatever small amounts of time, space and energy can be mustered. As the individual chapters detail, these are the stolen moments of learning English by the TV, dictionary in hand, late at night just before succumbing to sleep, reading on the bus to work or on the toilet at home, preparing supper while talking over work issues on the telephone, or finding time to slip away for an hour to the local library to peruse new garment patterns in peace. Complex and fragmented relations of paid work and home/family care characterize women's learning much more than they do men's. Men rely on women's greater unpaid work to find much more discretionary time for learning. As Luttrell notes in her research on working-class women,

> working-class women feel a deep conflict between self and others, placing their needs last either by choice or by force. Therefore, if learning is to engage working-class women, it must be presented not only as an indi-

271

vidual self-development process but as one that is rooted in family and community relationships. (Luttrell, 1997: 175)

Learning in paid workplaces is also intimately related to one's job. The operation of promotional and hiring patterns greatly dictates workers' positions and can therefore also be seen as a significant barrier to women workers' learning as well as reproductive of patriarchal order. To see this most clearly requires the type of expansion of the notion of learning beyond the formalization, individualization and content barriers that we have discussed. CHAT helps us see that learning is inextricably linked with forms of experience and activity that are formally defined in the workplace through job categories, organizational forms and managerial practice. These practices attempt to structure learning by limiting discretionary control over movement, scheduling and experience generally. This structuring effect is seen most clearly within the chapters in three contexts: the effects of formal job categories on learning (introduced above), the gendered pathways to new job categories, and the radical differences in learning patterns between trades/maintenance workers (virtually all male) versus general operatives (primarily female).

As the small parts, garment worker and college chapters indicated so clearly, women workers' access to positions with higher imputed skills (i.e., learning opportunities) is much more limited when compared to men of comparable background in the workplace. In the case of the small parts workers, female workers who were formally well qualified for apprenticeship and other training positions were effectively discouraged. A male worker in the same plant used what amounts to an informal apprenticeship[1] to learn the requisite skills and knowledge from another male, after which the worker (often vouched for by his "teacher") could make a convincing case for promotion when a position opened up. For the vast majority of female workers without formal credentials, the situation was even worse. Women in the college described the distribution of discretionary learning time by describing their work as "putting out fires" whereas the men draw work assignments off a "waiting list."

The division is more stark among the garment workers. Male cutters such as David who learned his "cutting" trade informally on-the-job, or Tony who has extra access to computer-training and even regular interaction with the design department, have plenty of opportunities to train. In contrast, the predominantly female home-workers are shut out almost completely from collective informal training, are poorly situated for formal training, and limitations in time, space and energy sharply structure their own informal learning. Perhaps the most radical demonstration of gendered barriers within working-class learning, however, appears in those chapters in which workers seldom mention "gender" at all. (Of course, this does not mean the barriers do not exist. Rather it points to the relative absence of resistance.) In the auto assembly and the chemical plants we researched, hiring and recruitment practices produced a 14:1 and a remarkable 72:1 male to female ratio on the shop floor.

Though affecting men to a degree as well, a key gender issue is the vastly different learning patterns that exist between trade/maintenance workers and the rest of the workers. Men account for virtually all the trades and maintenance positions in the workplaces we researched while the women were largely concentrated in the production positions, which are generally imputed to be less skilled, less well paid and are more likely to be part-time and/or contingent on just-in-time production demands. The effects of learning here are that skilled trades and maintenance workers generally exercise considerable discretion over the scheduling, pace, re-sources, learning and experience within their work. The contrasts between the small parts and garment workers were striking. On the one hand there were (predominantly female) small parts production operators who were demanding the opportunity to learn about production beyond their own tightly defined position and struggling to obtain the discretion to deter-mine when their own bladders were full (i.e., bathroom breaks). On the other hand were the male trades/maintenance workers in the plant who could schedule much of their own work, converse freely in a semi-private tool-room, and generally win control from supervisors with technical ex-pertise and language developed largely through their own discretionary activities throughout the plant and prior training. The garment industry re-vealed similar contrasts between the female home-workers and male fac-tory cutters. The women home-workers juggle childcare, machine mainte-nance and production tasks in stark, often competitive, isolation from each other at piece-rates often below minimum wages, while the predominantly male cutters in the men's apparel factories operate with greater wages, relatively greater job security and have the opportunity to consider techni-cal skill upgrading to CAD/CAM systems of apparel design. These trades/ maintenance positions allowed most direct control over time and space in which to engage in either individual or collective learning on the job.

Gendered mediation of learning is assessed here by comparing job-based informal learning patterns (Figure 9.2). Gender effects on informal learning are most likely to be found here because this sphere entails the direct mediating effects of occupation, firm, sector and union strength, and workers generally can exercise the lowest relative control, in comparison to home, community and other general interest-based informal learning (see Livingstone, 2002). The basic pattern confirms that men typically indicate higher levels of engagement in job-based informal learning, that is, "expansive" as opposed to degenerative, reproductive or "contracted" (Engeström, 2000). Garment workers who are predominantly women have very little discretionary time to devote to informal learning. Women average barely an hour a week while male garment workers have more than twice as much "free" learning time. College, chemical and small parts workers exhibit similar gender patterns on higher general averages. The auto workers, who are almost exclusively male, have much higher rates of informal learning than any other group of male or female workers.

Our analysis of home-based informal learning by gender finds a smaller margin of difference between male and female workers generally. But overall, even in non-employment spheres where workers have most control over their own affairs, patriarchal structures persist and women workers tend to have substantially less free learning time than male workers.

Figure 9.2: Gender Effects on Job-Based Informal Learning* Across Sites

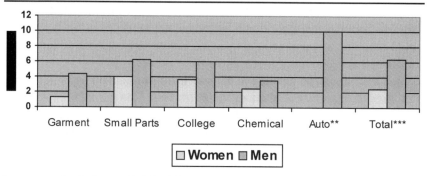

* responses begin from a 1hr/wk minimum; reports capped at 20 hrs/wk.
** only one woman interviewed in auto so her response removed.
*** Totals represents weighted averages based on numbers interviewed at each site.

Race/Ethnic Dimensions of Working-Class Learning

Our designations of race/ethnicity are based on interviewees' self-descriptions of their racial/ethnic background (collapsed into simple "white" and "non-white" categories because of limited numbers). Overall comparisons of race/ethnicity effects on job-based informal learning levels across sites show a more uneven pattern than gender effects, made less informative by the fact that reliable hourly levels could not be established for the very small numbers of "non-white" interviewees in the almost exclusively white auto site. Nevertheless, overall, white workers do report higher levels of job-based informal learning. A host of factors come into play here, however, and language barriers (despite our efforts to interview in participant's first language) may limit the validity of responses.

Figure 9.3 summarizes the basic patterns by the simple white/non-white dichotomy. Garment workers indicate the lowest levels of job-based informal learning overall. The majority of the garment sample includes recent (white *and* non-white) immigrant workers, with the distinction between white and non-white workers muted. Non-white workers in the chemical sector indicate very low relative amounts of job-based informal learning time – less than two hours per week, about half as much as white chemical workers. White and non-white workers in the small parts and college sectors, where white and non-white numbers are most equal, report roughly equal amounts of job-based informal learning time. White auto workers, in an almost exclusively white milieu, have by far the greatest amount of time to devote to job-based informal learning. The overall pattern is that whites are able to devote more than twice as much time per week to job-based informal learning as non-white workers are.

Figure 9.3: Self-identified Race/Ethnicity by Job-based Informal Learning* Across Sites (average hours per week)

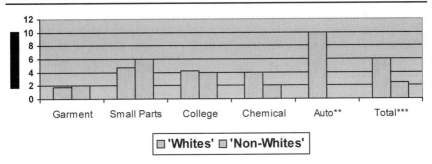

* Responses begin from a 1hr/wk minimum; reports capped at 20 hrs/wk.
** Only one "non-white" interviewed for auto so not included in chart.
*** Total represents weighted averages based on numbers interviewed at each site.

Despite measurement limitations, evident race/ethnicity differences appeared across the sites. The central forms through which these relations became the foci of contestation were in ESL policies and practices, assessment/recognition of foreign credentials and the deeply racialized segmented labour market referred to earlier, as well as (formal and informal) access to promotion at work.

Second language learning for adults is a difficult task even in the best of circumstances and its relegation for immigrant workers to attentive listening in the elevator, late nights by the television and careful reading of the newspaper or brief, interspersed encounters with co-workers makes advancement slow. What is more, low-level English literacy skills usually eliminate these workers from job categories that allow greater discretion over learning on the job. The inequitable assessment of foreign credentials is a quite specific mechanism through which capitalism is grounded in racialized relations. Underemployment of qualifications is particularly widespread amongst immigrant garment workers – this issue is closely connected with ESL, as workers with university and professional degrees are denied access to jobs for which they are otherwise well qualified. And, neither do the guidelines of immigration sponsorship help the situation. As garment workers sponsoring family members indicated, the rules of sponsorship lock the sponsored person out of social services. Situations such as this are not good for other workers in the sector either, as the extremely vulnerable position of immigrant garment workers increases downward pressure on wages in general. Several immigrant women home-workers described in graphic detail, the enormous pressures of contingent employment, below-minimum wage levels and a struggle to keep food on the table. In this context, working-class learning is most severely limited. Finally, the segmented labour market is constructed through job ghettos that are clearly split along racialized lines. On the most basic level, immigrant minority workers' employment in peripheral sectors deeply affects discretionary time, space and energy available for learning, through their lower wages, poor conditions and more contingent employment. Making matters worse, the strictly adaptive managerial practice and firm strategies in many of their paid workplaces entails even greater work intensification and constricted opportunities for workers. Such segmented workforces are evident among the line-working women in small parts and the custodial workers at the college as well as in the garment sector. Racial ghettoization in the workplace appears, on the basis of our evidence, to pose a significant challenge to the collective potential for working-class learning.

The interaction of race and gender effects is evident at several of our sites. Most obviously, in terms of relative power, in the auto plant where white, middle-aged white men dominate the workforce, we also tend to see the most intensive use of job-based informal learning as well as of training courses. Conversely, in the garment sector where non-white women predominate, both job-based informal learning opportunities and training courses are least frequent. However, the case of the small parts workers is especially instructive in terms of a more positive interaction of ethno-linguistic and gender effects. A campaign by immigrant women of colour for ESL courses at the factory during collective bargaining produced both formal and informal learning opportunities in spite of persistent barriers. A major barrier to progress has been the company's refusal to contribute paid time off for training (even when matched by equal contributions of workers' own free time) or fees to contribute to instructors'. Nonetheless, in addition to their informal language learning efforts, every Monday after their shift, many tired, hurried women have asked their families and husbands to make do without them while they show up for ESL courses in the company cafeteria. Inspiring indeed is the collective struggle of these immigrant women of colour for better learning opportunities.

Age Dimensions of Working-Class Learning

The most consistent empirical finding in all research on adult learning has been a strong inverse relation between course participation and ageing (see Courtney, 1992). The transition from adolescence to adulthood has always demanded a great deal of informal learning and a credentials-based labour market now also demands increasing formal schooling and further certification. Routinization of life patterns and accumulation of knowledge have generally been expected to diminish interest in pursuing further courses in later stages of the life course. Recent national surveys have reconfirmed these general patterns in terms of course participation (see Livingstone, Raykov and Stowe, 2002). In addition, unionized workers of all ages have higher general participation rates than do non-unionized workers, reflecting in part their greater power to negotiate employer support for course provisions in relation to changing job requirements (Sawchuk, in press). But there is little doubt that older workers have less interest in committing substantial time to course participation. The common assumption has been that older workers have less interest in job-related learning generally.

However, recent surveys inclusive of all forms of organized training – including workshops of short duration – have found that the majority of employees, including many older workers, are now participating annually in some forms of organized retraining, often associated with enterprise restructuring or technological change (Livingstone, 2002). As we saw in the site chapters, unionized workers of all ages have at least modest continuing interests in at least limited continuing studies in relation to their jobs or employment conditions. Even more significantly, informal learning has recently been found to persist among most social groups at a high level through the later stages of the life course. The first Canadian national survey of informal learning conducted in 1998 (see Livingstone, 1999b) found that, while very young workers had much higher rates of job-related informal learning, older worker had only slightly lower rates than middle-aged ones. Figure 9.4 summarizes the incidence of job-related informal learning by younger and older workers at the five sites.

Figure 9.4: Age Effects on Informal Job-related Learning* Across Sites (average hours per week)

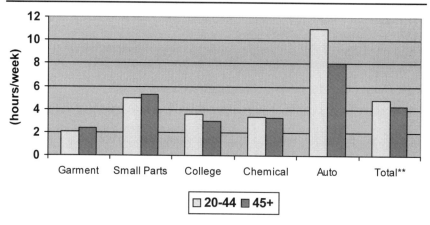

* Responses begin from a 1hr/wk minimum; reports capped at 20 hrs/wk.
** Totals represents weighted averages based on numbers interviewed at each site.

The aggregate pattern of job-related informal learning across all five sites is broadly consistent with recent national survey results, with slightly lower intensity of informal learning by older workers. But the inter-site patterns are more explicable in terms of specific sectoral features, differences in union strength and seniority privileges. The lowest rates of job-based informal learning among both younger and older workers are in the garment sector with the most fragmented production sites, least union strength and a predominance of precariously employed younger women workers. In fact, older garment workers tend to have slightly higher rates because of the higher incidence of older trades workers with greater bargaining power. While the small parts plant union is also relatively weak, recent managerial and technological changes have stimulated renewed informal learning, especially among more senior skilled trades workers. The college and chemical sites, which have similar overall incidence of job-related informal learning and the auto plant, which has the highest incidence, all exhibit the dominant pattern of slight decline in job-related informal learning with ageing. But the most notable pattern here, in contrast with the widespread stereotype of declining interest in learning amongst older people, is that older workers at all sites remain active learners in relation to their employment and that any declines are really quite slight. This pattern may be partially explicable in terms of the mitigation of any negative aging effects by the greater security and discretionary control of older unionized workers by dint of their seniority privileges but, in general, older organized workers continue to be active informal learners.

Both the site studies and these comparative analyses suggest that simple linear predictions of age-based declines in workers' learning practices are inaccurate. Many other factors influence the incidence of workers' learning through the life course. But it is reasonably clear that workers in stronger unions, able to fight for more extensive training provisions, are more likely to continue to have more formal educational opportunities and be enabled to spend more time in relevant informal learning through their job lives.

Of course, all of the above findings on gender-, ethno-linguistic- and age-based differences are very preliminary, given the small exploratory samples involved. But once more, the utility of applying a knowledge/power perspective to a CHAT approach to understanding adult workers' learning practices appears to be demonstrated.

Recommendations

In terms of recommendations from this study it is helpful to first explore studies which provide a comparative backdrop to our approach as a whole. From here we can then investigate the most important and most practical recommendations from the sites themselves.

Comparing Findings and Recommendations of Related Studies

We will put our recommendations in comparative perspective through a discussion of the recommendations of others who have also looked carefully at issues of workers' learning. For our purposes this entails a brief but focused review of two contemporary large-scale empirical studies, one based in Canada (Betcherman, Leckie and McMullen, 1997), and another based in the United Kingdom (Forrester, Payne and Ward, 1993; 1994; 1995). These other large-scale empirical studies of workers' learning appear to have some limited correspondence to the outcomes of our research. There are some good reasons for this. A key issue to be understood in comparing these studies is the breadth or, more precisely, the narrowness of conceptualization of workers' learning. We have chosen to think of learning very expansively and this makes comparative analysis more difficult.

Most comparable to our study is the *Learning at work. Final report. Leeds Adult Learners at Work Project* (Forrester, Payne and Ward, 1993), which is also discussed within *Workplace Learning: Perspectives on Education, Training and Work* (Forrester, Payne and Ward, 1995). The differences between our project and the British project revolve around issues of focus, purpose, theoretical approach and vantage point, as well as the specific national and industrial relations contexts of the research. Despite differences, the projects nevertheless intersect at the point of the need to more fully understand workers' learning. In the final chapter of Forrester et. al (1995), the authors reiterate a summary of their recommendations in terms of the following (taken from Forrester *et al.*, 1995: 156–157):

1. Networking and partnership are important means of generating new Employee Development schemes, but in the longer term a National Employee Development Agency should be established as a focus for information, consultancy and evaluation.

2. The cost of Employee Development schemes should relate to clearly established priorities and offer the same entitlement to those within and outside of paid work.

3. Employee Development schemes should be constructed so as to maximise personal growth opportunities for employees.

4. There is a common interest for employers, employees and government in developing a comprehensive and transparent system for accrediting workplace learning.

5. Lifelong learning provides a context within which "a learning culture: in the broadest sense can grow and flourish in the workplace."

6. Most small and medium-sized enterprises (SMEs) need external support in developing Employee Development schemes.

7. There is an urgent need to develop models for the successful organisation of Employee Development schemes.

8. Given the complexity of adult motivation to learn, it is important to ensure that independent guidance is available to employees within Employee Development schemes.

9. A key role of the proposed National Educational Development Agency would be to research and monitor Employee Development.

Forrester *et al.* present interesting macro policy options, though they do not, in our view, make a sufficient attempt to understand already existing learning practices in which workers are engaged. Informal learning, in particular, is insufficiently recognized. This produces a tendency, in the reading if not the intentions of the authors, to see the problem of work-based learning as a problem credentializing learning rather than reorganizing work. Our analysis demonstrates that these problems are an expression of local, sectoral, national (and possibly international) structures that inhibit the full realization of workers' current learning capacities. As we have outlined, workers' learning is currently very expansive, with the central problems of learning in the workplace and beyond rooted in class, ethnic/racial and gender relations under capitalism. The research of Forrester *et al.* is more so a study of learning for and within capitalism per se than it is a study of workers learning within *and beyond* capitalism. One of the other obvious reasons for the differences in our respective recommendations relates to Forrester's *et al.* engagement with very specific forms of the British workplace training policy debate of the 1990s. As they indicated, their work is largely targeted at the level of policy. However, in not providing an account of the full range of actual learning practices, their specific policy recommendations are based on partial information and thus limited in some ways.

It also appears that the recommendations of Forrester *et al.* (1995) are also based on their desire for middle ground, i.e., "only by re-establishing the dialogues within the economy between the "social partners," and between economy, community and society, can individuals find any long-term prospect of a renewed sense of order in their lives" (150–151). This "middle ground" rests amidst the tensions of paid work within which they call for policy-makers to balance the needs of workers and employers. "[E]mployee needs for personal development (social needs) must be set alongside employer needs for higher level skills (economic needs) if real progress is to be made in developing the whole workforce" (111).

Forrester *et al.* take up the notion of "Employee Development" (ED) as their central rallying point and set themselves the task of balancing the needs of workers and employers. Subsequently, they recognize a significant contradiction:

> Throughout the empirical research that we have carried out, there remains a constant tension between, on the one hand, those training needs identified by employers and the felt needs of employees for personal development. (Forrester *et al.*, 1995: 110)

Their general solution to the tensions of work is to argue for broad-based education as opposed to narrow job-specific skill training to help Britain better compete in the globalized marketplace. Such programs, it is argued, would balance the needs of the employer (to help employees better understand the organization, encourage flexibility, develop a learning culture, and create an internal labour market) with the possible gains for the workers (compensation for negative experiences in earlier education, elimination of barriers to participation caused by shift work, employee development beyond current job requirements, equal opportunities, personal confidence and preparation for an unpredictable labour market) (Forrester *et al*. 1993). The outcomes for workers that Forrester *et al*. suggest as part of expansive ED programs overlap with some of our more specific "Work-based Learning Initiatives" recommendations below.

But, we believe it is premature to prioritize balancing the needs of the wealthy and powerful with those of the poor and relatively powerless until the latter's needs are clearly understood by themselves and others. Our view is that for any social research that aspires to contribute to actual social change, not only are rigorous and systematic methods of social research required, but researchers must make explicit their political standpoint. The political commitment to "the social goal of lifelong learning" is not sufficient, in our view, to move distinctively and decisively beyond employer perspectives. Given the progressive research and educational tradition from which the Leeds project emerges, undoubtedly the authors struggled with these very issues. In the context of the fact that knowledge must be situated in its multiple sites of production, intellectuals must decide whose knowledge claims are most in line with their underlying political project. Our research process explicitly chose to pursue the knowledge claims of workers and, in this sense, attempted to explicitly operate from the vantage point of workers.

In comparison to the Leeds study, the major Canadian study from Betcherman *et al*. (1997), *Developing Skills in the Canadian Workplace: The Results of the Ekos Workplace Training Survey*, is more narrow. Its goal is to understand and improve the efficiency of Canadian companies and their training programs. The audience and purpose are clearly those of large- and small-scale capital and the authors expend little effort trying to convince people of their neutrality. The Betcherman *et al*. study does, however, have some important lessons to teach us, some of which amplify some of our own findings and recommendations. Briefly, their recommendations are broken into three sets of problems: "polarization," "small firms," and "individuals without access to training." The first problem deals with simply identifying the disparity of organized training operations across different types of firms and sectors, suggesting that firms should do what is best suited to their situation. Large, technologically advanced firms should expand their training operations, and those firms with small operations and/or which operate with "traditional technologies" should find a more creative approach to do the same thing. The second problem focuses on how small and/or disadvantaged firms should appeal to community councils, sectoral councils and/or the government educational institutions. Access to these structures could be most efficiently managed, according to the authors, through the establishment of networks that "work toward building relationships among the business community" (xii). This is essentially a recommendation to socialize the costs of training while retaining private control over the gains or profits. The third problem is of more interest to us, though it requires further interpretation. As in the previous problems, for Betcherman *et al.*, there is the presumption of the "human capital" model that we criticized in the Introduction and Chapter 2.

The Betcherman *et al.* research focus leads towards very different choices around data collection, interpretation of data and conclusions. For these authors the focus is on learning for individual wage gains and company profitability, and the research's vantage point is that of capital first and foremost. Given this, the emphasis is on creating more competitive capital nationally by more fully enlisting the creative powers of skill and learning among workers for use by capital. The short quoted statement in the Executive Summary of Betcherman *et al.*'s (1997) report (xiii) is at least quite clear in terms of its standpoint: the commitment revolves around hierarchical organization of decision-making in a firm (i.e., it is only critical for managers to be convinced of the worth of particular workplace learning strategies), and workers as sellers of labour rather than active agents of creative change. However, the commitments, conclusions and recommendations tend to undermine the methodological choices and the actual data although, to be clear, their empirical findings (not their recommendations for concrete change), *are* consistent with our findings in terms of the dominance of informal learning, as well as the massive gendered, "skill," wage, educational attainment, age and (via second-language status) racialized biases in workers' formal training. Methodologically, the authors inform us of their lack of control over survey distribution among workers, lack of resolution of the problem of accessing the perspectives of employees with "literacy problems" (18) and, most importantly, as in Forrester *et al.* (1995), the reliance on employer-selected worker-interviewees.

> Given these possible sources of bias in the employee survey, it is clear that the survey cannot be regarded as being representative of all employees in the case study establishments.... These kinds of problems have beset most previous attempts to develop linked employer-employee survey

methodologies. Undoubtedly, they explain the fact that this type of research has rarely been attempted on this scale. The state of the research is still fundamentally that of testing methodologies. (Betcherman *et al.*,1997: 18–19)

The problem is less a matter of methodological choices and research practices than the lack of recognition of diverse social standpoints. In the case of Betcherman *et al.*, the authors undertake a "linked employer-employee survey" methodology and, despite useful empirical work, filter workers' perspectives through a managerial standpoint, rather than allowing workers' standpoints, interests and world view to emerge in their own right.

WCLS Project Recommendations

As we indicated in Chapter 1, in each work site of the WCLS project, preliminary reports on findings and recommendations on working-class learning strategies in the workplace, home and community were presented back to the membership in various ways. These results were inserted into a consultation and revision process that culminated in a closing forum of representatives and researchers from each research site. The recommendations were then refined and final reports were given to the locals for their own use in internal educational program development and negotiations with employers and public agencies. Before discussing the specifics of these recommendations it is worthwhile to hear briefly from a few of the union representatives themselves on some of the most important basic themes that the research helped surface in their locals.

> The initial report was very well received in the local ... The results of this clearly showed that the workers I represent, which are the same workers that others in the room represent, are not the stereotypical factory worker ... I mean the talent is there, it's unbelievable – or is it? It is not really. It comes as a surprise only if you underestimate workers ... (Chemical Workers' Union Local Representative).

> What surprised me is, I did not realize it was the same for everybody else, that even though we're in an educational institution, there's no real on the job training at the College. No one is ever there to show you how to do your job. You're simply expected to pick it up and run with it ... So, overall, I think, the lack of training to do your job seems to be the most outstanding surprising thing ... I became president partly because of this study, because of some of the things that we were upset about in our workplace when we read the study, and that was that our local union was training people itself without any credit ... And we got some more stuff in our collective agreement about job training, partially through this study

and partially through other people who've shared their experiences. You know, that's the way to learn. (College Staff Union Local Representative).

Actually, once the programs we recommended through the study were put into place, like it was assumed that only something like three people would sign up, but pretty well everybody signed up. It was amazing, you know, once they were given that opportunity. Prior to that opportunity, like I said, everybody, oh, well, nobody will come, nobody will come. And, like, we'd have five pages of names signed with people wanting to learn. (Small Parts Union Local Representative).

Against this background theme of enormous learning capacity among workers we can turn to a comparative review of the recommendations across sites as a practical guide for the further development of efforts at directly developing greater opportunities for workers.

Table 9.4 compares the main recommendations arising from the different sites using two categories: "Job-Based Learning Initiatives" and "Broad-Based Learning Initiatives." Organizing the recommendations in this way helps us compare our recommendations to those of other researchers, and also continues to develop upon the broad orientation to "learning" which is distinctive to our project. The rare prior empirical studies of working peoples' learning tend to deal with workers' workplace learning in a narrow way which defines relevant learning activities and perspectives according to dominant needs of the capitalist economy if not the local employer specifically. In most of these studies, conceptions of learning are bound by the desire for competitive advantage for business as well as, under the rubric of human capital theory, individualized labour-sellers.

Oriented to the practical and more or less immediate needs of *particular local* sites, the recommendations of the Job-Based category are important but quite basic. On their own, they offer only limited opportunity for enhancing practice. The recommendations of our Broad-Based category, however, suggest some provocative openings for new forms of support for working-class learning. These are less bound by the logic of production. Rather they are directed at expanding upon wide-ranging interests, expanding opportunities for new forms of participation that may allow workers to use and develop more of their skills and judgement. Moreover, these recommendations recognize that working people want to, and in many cases do, pursue learning beyond the confines of paid work. The items in the Broad-Based category collectively paint a picture of working-class learning grounded in the realities of the day (e.g., demand for more computer skills), and recognize the needs of everyday practice in the paid

Table 9.4: Summary of Site Recommendations

	CAW	CEP	OPSEU	USWA	UNITE
Broad-Based Learning Initiatives					
Computer Access / Training	—	X	X	X	X
ESL Training	—	X	X	X	—
PEL	—	—	X		
Broadening Non-job Learning Networks	X	X			
PLAR Program	X	X			
Facilitation of Non-formal, Non-job Learning	X	X			
Further Research		X			
Basic Literacy Programs	—		X	X	
Newsletter (as tool for learning)	—	—	—	X	
Job-Based Learning Initiatives					
Need for Collective Bargaining on Training	X	X	X	X	X
More Opportunities for Informal Learning from Co-Workers	X	X	X	X	X
Reorganization of Training Structures		X	X	X	X
Reorganization of Labour Process		X	X	X	
(More) Joint Programs	X	X	—	X	
Pilot Training Programs		X	X	X	
Workers-as-Educators Programs	—	X	X	X	
Local Union Education Officer	X	X			
Courses to Address Learning Barriers at Work			X		
More Union Courses			X		
More Courses on the Production Process	—	—	X		

Legend: X = Specific recommendations by workers interviewed; — = notions discussed by workers but already in operation in the workplace; "blank" = not discussed by workers.

workplace as well as home and community. These proposals also respond the commonly expressed desire to understand the paid workplace beyond one's own, sometimes quite isolated, work station (e.g., learning about the larger production process).

These recommendations are also based on the desire to have greater opportunity to explore formalized learning (e.g., PEL, PLAR) with greater access to further non-employment–related learning. Perhaps the most important distinction however is that some of the most important learning that workers undertake involves "networks" within the different everyday communities, e.g., broadening non-job learning networks, workers' newsletters.

Notions of the employers' obligation to support and compensate work-based training underlies virtually all the recommendations from our own research sites. The importance of specific items including PEL, the elimination of inequitable training and learning structures in the workplace and beyond, and access to basic literacy and numeracy upgrading are all asserted in our local research sites, as is the reaffirmation of the central importance of union-based education for members. At the same time, our research adds considerable empirical focus to these broad statements. The details of achieving the above principles necessarily vary from site to site. Different union traditions, demographics, locations, sectors, organizations, size and overall situation demand this. While overall strategy and goals remain the same, local priorities and tactics shift.

Beyond the call for such specific programs as ESL, Basic Literacy/ Numeracy (in some cases), computer courses and local Workers' Education Centres, a comparative look at the recommendations that appeared across our sites revealed four key interrelated features:

1. A general call for an increase in worker mediation of education, training and learning structures of all kinds.

2. A call for the formalization and accountability of changes through collective bargaining.

3. A call for the facilitation of broad-based home- and community- as well as workplace-centred initiatives.

4. A call for the recognition of, compensation for and strategizing around informal learning.

The most common, foundational feature that the site recommendations shared focused on the call for a general increase in worker mediation of work and learning. These recommendations pertained not only to specific programs and committees in which workers should be directly involved, but also to such duties as curriculum development and delivery of programs. They also highlighted the need for worker mediation over structures and decisions that less obviously affected learning and which have traditionally (and legally) been seen to be the prerogative of management, such as job and skill categories, shift scheduling and even the broader labour process itself. As the site chapters clearly point out, traditional training issues and initiatives are only the tip of the worker learning iceberg.

The auto assembly chapter offered one of the more developed, practical visions of this call for greater worker mediation. Through a long history of militancy and collective struggle, the CAW local had become fertile ground for moving forward in ever broader learning initiatives. More than the others, recommendations from this site seemed to jump directly to practical discussions of gains that can be made through the collective bargaining model and changes within the union local. Joint programs that enlist the resources of the company, the explicit promotion of education as a workplace right, PLAR to facilitate learning in and beyond the workplace – all administered by a Union "Educational Officer" – were among the suggestions from this site. This level of strategizing begins from an already well-developed base of hard-won gains affecting informal and formal training, formalized skill development and promotional structures (closely related to the seniority system) and extensive union-based programs such as PEL.

The chemical chapter offered another perspective on the call for greater worker mediation, but one that began from a very different workplace structure which included the company's use of the OJT structure as well as formal training. Workers in this site, partially in the course of this research, have come to take very seriously the informal learning upon which the OJT system is based. They also put much energy into discussing alternative systems of health and safety representation, an essential part of an important network of learning in the workplace. As one worker blurted out in a focus group session: "If the worker does not train the worker – who will? Consultants? It'll cost thousands!" Recommendations in this site were directed towards winning more resources and support from the company in order to develop the learning structures that are not only important to production, but are also the foundation of opportunities for more advanced job skills and career development and, as we have noted, healthy and safe work.

The context of the calls for greater worker mediation among the college workers was again quite distinct. In the midst of government restructuring and downsizing, this environment was saturated with uncertainty and stress and a dizzying array (at least on paper) of training, evaluation and participation structures (including: a "Centre for Equity and Human Rights," a "Centre for Employee Development," a "Participatory Management" program) and the local's own initiative to develop their own job and "skill sets" profile to understand and build greater worker involvement. These formal structures notwithstanding, the recommendations chronicled the need to break down hierarchies of learning opportunity that emerge from employer-defined job and skill categories which intensified deep divisions between workers along ethnic, racial and gendered lines. Here the notion of a "job" versus a "career," and the possibility for the meaningful development and use of expanded work skills shaped the specific types of recommendations put forth. Central among the recommendations were the ideas of putting the rhetoric of "Participatory Management" to the test, and seriously developing the "Workers as Educators' initiative.

"Workers as Educators" and greater control over key structures of learning were also central features of the small parts site recommendations. Here barriers to learning from the labour process itself were highlighted. In close connection to this, promotional structures were also identified as part of the problem of equitable distribution and expanded opportunity for workers' learning. Local workers and the union executive used the momentum of the findings of this research, in fact, to help establish the workplace's first Joint Educational Committee to guide the development of a Workers' Education Centre which was successfully bargained for and which remains in operation (though at a different facility). However, increased worker mediation did not stop there. Other recommendations dealt with core skills (ESL, basic literacy and numeracy, computer skills) that would add to the overall capacity for more meaningful participation of workers in learning activity within and beyond the factory gate. There was an overall questioning of the structure of training in this workplace which was closely tied to (formal and informal) promotion and "career-path" structures. Much like the college site, certain groups seemed stuck in more limited job and skill categories (these were predominantly immigrant women, ESL workers on the line), whereas others had more discretionary time and freedom of movement within production, with expanded, meaningful opportunities for both learning and advancement (white, male, skilled and semi-skilled workers). In this sense, the structure of the work process itself was critiqued for its limitation of meaningful learning opportunities along the lines of skill categories which were seen to be race- and gender-based.

The recommendations arising from the garment workers' interviews reiterate how, in yet another context, workers were engaged in a struggle for greater forms of participation in the work process. The terms of this struggle were structures of learning in the paid workplace as workers and beyond it as citizens from non-English speaking countries facing types of isolation in a secondary labour market. The structures of work and training were heavily critiqued. Indeed, the degree to which training structures were absent was highlighted. Here, employers were dependent on the import of skills rather than developing their own means to reproduce worker knowledge and skill. Employers' attempts to develop sweatshops and home-based piece work further reduced chances for workers to get together and learn informally. Job or career progression, so closely tied to gaining and using new skills, was also largely absent, aggravated further by the isolation and demanding context of home-work in the sector. However, despite difficulties, a strong underlying foundation to recommendations indicated that gains should be formalized within the collective bargaining process (and extended to those denied access to one). As in the small parts site, overall capacity to gain greater control over their work and lives was also seen to be connected to developing the more foundational skills such as ESL and computer literacy. These foundational recommendations were seen as essential to "kick-start" participation and were central to the development of other measures.

Workers across the sites also recognized tools for change beyond collectively bargained measures. There were recommendations for things that supported learning and solidarity in a more informal, everyday way through the use of worker newsletters as an educational tool, the broadening of non-job related–learning networks and the need for more union-based research on educational issues. However, most of the fundamental recommendations in each site were more or less directly rooted in the collective bargaining process. The auto assembly site focused on the development of joint structures, PLAR, etc. The chemical site focused on developing language to gain more resources for the OJT worker-instructors, and on negotiating a formal resolution to the unofficial "rover" position upon which opportunities for internal apprenticeships and further learning seemed to rest. College workers called for contractual recognition of, and compensation for, the expansive training they do on their own, as well as for practical means of genuine involvement in the declared participative structures. The small parts workers were the swiftest of all, to the credit of both local and district union support, immediately negotiating their on-site "Workers' Training Centre" which generated momentum for several other recommendations, despite the relocation of the plant. The garment worker recommendations focused on needs for foundational courses and structures such as a resource centre. These were union-centred initiatives of the two merging unions, but they also pursued the collective bargaining process and joint initiatives with remaining large employers and governments in a rapidly fragmenting industry.

Our description of the auto assembly site is perhaps the clearest suggestion of how strong training and educational contract language can help guard learning from being, in effect, "colonized" by competitive internal and external labour markets. Our findings show a resistance by workers to an "instrumentalized" approach to learning. In fact, the recommendations of both of the most strongly organized sites culminated in the recommendation for some sort of local union education officer to translate new recommendations into both the collective bargaining process and the wider community, and ultimately to help organize their development and enforcement over time. Other sites found themselves dealing with more troubling situations. Sites with a less developed culture of union learning needed to build basic awareness of barriers. In the absence of programs like PEL and general educational tuition support, they needed to build greater foundational capacities in areas of ESL and basic skills for all members, and generate some of the more basic means to provide their membership with opportunities to build momentum for more effective contract-based discussions. Of course, the sectoral and organizational features inherent in each site play a major role here. It is more difficult to establish rich union cultures under conditions of isolated work and in the face of constant cycles of "contingency", lay-offs, restructuring and downsizing.

One unique feature of this research and our site recommendations was the refusal to limit the scope of working-class learning. This is, as we said, partially due to our commitment to the vantage point of working people rather that of employers. However, it is also rooted in our theoretical critique of dominant educational forms that, we maintain, delegitimize much working-class learning. The broad-based recommendations for expanded, non-job learning are partial testimony to this. PLAR and PEL are the two most readily available tools to allow workers to realize their learning goals in the realm of further education. However, other recommendations for the expansion of non-employment informal learning networks were also made in several sites. It should be reiterated that workers are not only interested in positioning themselves in the labour market through extra credit and skill acquisition. In the first place, they have described the many ways in which they want to be able to take a more active, creative role in their work so that they are more than cogs in a machine. *This* is the core reason why skill acquisition must be accompanied by increased opportunities to use new skills. Existing capitalist organizational structures in this research were seen to be far too restrictive specifically in terms of skill categories and promotional opportunities, i.e., the construction of "careers" versus narrow "jobs." However, even more directly in terms of broadness, working people in our research outlined how they want to use learning to build their own communities beyond the workplace and add to the quality of cultural life in the home and neighbourhood. This rich and vast array of skills and knowledge, which is acquired and continually developed largely through informal means, is a feature of working-class cultural life which, as we argue in Chapter 2, is patently denied, dismissed or denigrated by most conventional education, training research and policy-making.

Finally, many of these recommendations were centred around the realization that working people's informal learning is of great value, to them, their families and friends, and indeed to their employers. Communities are in many ways sustained by the relations of knowledge and skill exchange and mutual aid. As many of the workers in this research came to understand, even in the seemingly most mundane of operations, economic production soon grinds to a halt when workers' informal learning, creative agency and cultural knowledge are withheld. The practical outcome of this realization is the need to develop specific strategies that can make progressive use of these findings. Recommendations for facilitation of informal, non-job–related learning are part of this growing realization. At the chemical site, workers have had intensive discussions over the structure of the health and safety information system and reaffirmed the importance of having an informal information network based upon departmental (in addition to centralized) representatives. They have continued to pursue this issue in both contract negotiations and internal reorganization of the local. In taking part in research aimed at developing a PLAR instrument at the local, workers made the suggestion that such an instrument, beyond its applicability to formal schooling and the labour market, could also be used to facilitate the development of the knowledge base of the local union as well as the informal networks of non-job–related learning (Sawchuk, 1998).

Other examples of the growing recognition of the centrality of informal learning include key recommendations of the college workers. Overloaded with change initiatives and shifting structures and context, they are doing more informal, "continuous learning" than ever before and are seeking to strategize organizationally around these issues. We have stressed the centrality of informal learning in building the strong union culture of the auto plant. Garment worker recommendations for a central resource centre are also based on an increasing belief in the ability of workers to control and administer their own learning formally and informally, given the basic resources.

Concluding Remarks

Our recommendations affirm the centrality of existing learning practices among working people and working-class organizations and the labour movement specifically. It is vital that researchers understand the foundational implications of issues of standpoint in the research process (Chapter 1); that organizational as well as larger political economic structures cannot stand unquestioned; and that workers, as creative agents in the workplace and beyond cannot be ignored within the construction of our research problems. Our research has tried to explore and document the actual learning practices of working people, and to develop policy and program suggestions in direct cooperation with workers as agents of their own political, economic and, ultimately, learning interests. One of the few writers to have taken a genuinely critical look at working-class learning explains:

> The skills of productive life are not limited to the workplace or to paid labour, but at the workplace it is workers who are training workers, not employers or "industry." It is only in dialogue with this existing practical knowledge that any technical progress takes place at all, let alone artistic and scientific development. However, without work, training is left in a social vacuum and skills transfer through culture atrophies and dies. When education attempts to fill this vacuum the bad experiences that many unemployed people have had with their previous academic schooling mean that many will not even consider a formal training option. They want jobs and may be desperate enough to take a job at any price but, like most people, what they would really like are jobs with prospects of progression and development that use their talents and abilities to the full. (Ainley, 1993: 47)

North American workers are marginally more likely to have had positive experiences in high schools and colleges than the British workers on whom this observation is based. More generally, we would insist that, given our broad understanding of "learning" as endemic to the human condition, it never ceases during our lifetimes. But we do nonetheless recognize the blatant hostilities, despite progressive rhetoric, of the capitalist labour process to recognizing and valuing much of workers' learning. Our recommendations seek to recognize the real barriers to meaningful workers' education through a dialogue with existing practical knowledge, as the above quote nicely articulates. We affirm that, at the grassroots, workers' learning can be aided by collective negotiation. Facilitating opportunities for more rewarding learning is part of a larger vision of sustaining and reaffirming working-class communities. In light of both the positive results we have obtained and the real danger of continued domination of the social research/policy agenda by (currently) more powerful interests, systematic critical study of the type we have attempted here should be continued, increasingly led by well-qualified labour movement researchers themselves. Such research may be pivotal to achieving educational equity and social justice for working people.

Notes

1. This does not include programs that are *strictly* "labour adjustment" initiatives which are abundant in the garment industry and at the college site we researched. Such programs are by their very nature transient and although important in their own right are not the primary focus here.

2. By non-formal apprenticeship we mean a learning experience that is structured (albeit loosely) but which carries a credential of value only within the company itself.

APPENDIX:

Interviewee Profiles

Worker	Age[*]	Sex	Job	Senority[†]	Race[‡]	Formal Schooling
AUTO						
Dick	30s	M	assembler	10	E.E.	some HS
Tim	30s	M	assembler	15	W.Can	some HS
Harry	40s	M	assembler	20	E.E.	college
Barney	30s	M	operator	10	W.Can	HS
Jill	40s	F	assembler	20	W.Can	college
Ronnie	40s	M	operator	15	W.Can	some HS
Mickey	40s	M	trades	15	W.Can	trade cert
Bennie	40s	M	operator	15	W.Can	some HS
Greg	40s	M	operator	20	W.Can	some HS
Sam	30s	M	lead hand	15	Asian	HS
Gerhard	60s	M	trades	20	W.E.	trade cert
Matt	40s	M	relief man	20	W.Can	HS
Darrel	30s	M	assembler	20	W.Can	HS
Pete	20s	M	assembler	10	W.Can	university
CHEMICAL						
Tom	30s	M	batch	5	W.Brit	some HS
Pete	30s	M	high perf	15	W.Can	some HS
Alan	40s	M	shipping	25	W.Can	some HS
Lee	30s	M	batch	1	Asian	HS+some univ.
John	40s	M	high perf	10	W.Can	HS
Teddy	30s	M	batch	1	W.Can	HS
Sean	30s	M	batch	1	W.Can	university
Enzo	30s	M	trades	5	Ital/Can	HS+trade+some univ
Jane	50s	F	batch	5	W.Can	some HS
Valerie	40s	F	batch	20	W.Can	some HS
Rick	40s	M	batch	5	W.Can	univers+prof
Dean	50s	M	batch	30	W.Can	some HS
Donovan	50s	M	batch	15	B.Can	some HS
Sam	20s	M	batch	5	W.Can	university

Worker	Age*	Sex	Job	Senority†	Race‡	Formal Schooling
CHEMICAL con't						
Sashi	40s	M	batch	10	Pak/Can	university
Dan	30s	M	batch	5	W.Can	univ+prof
Raymond	30s	M	batch	5	W.Can	HS
Lionel	40s	M	batch	10	B.Amer	some HS
Ron	40s	M	batch	10	W.Can	some HS
Vince	40s	M	batch	10	W.Can	some HS
COLLEGE						
Jackie	30s	F	clerk	10	W.Can	university+prof
Jocelyn	50s	F	nurse	30	W.Can	university
Lewis	30s	M	technician	15	W.E.	HS+tech dip
Florence	60s	F	clerk	20	W.Can	HS
Matt	30s	M	technician	10	Ital/Can	HS+tech dip
Ken	40s	M	custodian	20	B.Can	some HS
Lilly	30s	F	cleaner	1	Grk/Can	HS
Cathy	30s	F	clerk	5	Asian	HS
Frank	40s	M	maintenance	15	W.Can	HS+trade
Maureen	50s	F	clerk	15	W.Can	HS
Connie	40s	F	clerk	1	W.Can	HS
Norm	40s	M	admin	5	W.E.	university
Alice	30s	F	clerk	5	W.Can	university
Bob	30s	M	admin	5	W.Can	university
Sara	20s	F	admin	5	W.Can	university
Marco	50s	M	maintenance	10	Asian	trade
Alphonse	40s	M	admin	20	W.E.	university
Jack	40s	M	technician	20	Ital/Can	HS+tech dip
Lacy	40s	F	technician	20	W.Can	HS+college
Ellen	30s	F	clerk	10	Grk/Can	university
Lisa	40s	F	technician	10	W.Can	HS+some univ
Scott	30s	M	shipping	5	W.Can	some HS
Jean	20s	F	security	5	W.Can	HS
Ursula	50s	F	admin	5	Asian	university+grad sch
Worker	Age*	Sex	Job	Senority†	Race‡	Formal Schooling
Carrie	40s	F	clerk	N/A	W.Can	S+some college
Richard	30s	M	admin	15	W.Can	university+grad sch

Worker	Age*	Sex	Job	Senority†	Race‡	Formal Schooling
LIGHT MANUFACTURING						
Vlad	40s	M	main./trade	5	E.E.	tech dip+trade
Eduardo	—	M	set up	5	S.A.	trade
Jodi	30s	F	inspec.	5	W.Can	HS
Mary	40s	F	mat. handler	15	Ital/Can	jr HS
Lew	60s	M	main./trade	25	Ital/Can	primary
Rosa	60s	F	production	25	Ital/Can	primary
Marco	20s	M	mat. handler	5	S.A./Can	HS
Ricardo	30s	M	set up	15	S.A.	HS
Carlos	50s	M	main./trade	15	E.E.	tech dip+trade
Mike	40s	M	main./trade	15	W.Can	tech dip+trade
Gina	40s	F	production	15	E.E.	jr HS
Anne	30s	F	inspec.	15	W.Can	HS
Rava	20s	F	production	5	E.E.	HS
Daniel	60s	M	main./trade	15	S.A.	tech dip+trade
Frank	50s	M	main./trade	25	W.Can	jr HS
GARMENT						
Tony	40s	M	cutter	20	Asian	some HS
Sue	30s	F	sewer	10	W.Can	HS
Malcolm	50s	M	cutter	20	W.Can	HS
Pramila	20s	F	packer	10	SE.Asian	HS
Jerry	50s	M	cutter	20	Asian	HS
David	40s	M	cutter	25	Asian	primary
Lee	50s	M	cutter	-5	Asian	primary
Hien	50s	F	sewing	-1	Asian	college
Mei	30s	F	homeworker	N/A	Asian	primary
Sun	30s	F	homeworker	N/A	Asian	primary
Hong	40s	F	homeworker	N/A	Asian	some college
Hoa	30s	F	homeworker	N/A	Asian	college
Maria	50s	F	hole-maker	15	Port/Can	primary
Josie	20s	F	sewing	-5	S.E.Asian	HS
May	30s	F	packer	10	Asian	some university
Nympha	20s	F	sewing	-5	Asian	HS
Sina	30s	F	packer	-5	Asian	some university
Paula	50s	F	sewing	15	Ital/Can	primary

Worker	Age*	Sex	Job	Senority†	Race‡	Formal Schooling
GARMENT con't						
Joya	30s	F	sewing	15	Port/Can	jr HS
Donna	30s	F	sewing	15	B.Carr	HS
Kim	30s	F	homeworker	N/A	Asian	primary
Liz	30s	F	homeworker	N/A	Asian	HS
Lui	30s	F	homeworker	N/A	Asian	HS

Notes:

* Age has been rounded off for anonymity

† Seniority has been rounded off for anonymity. N/A=not applicable.

‡ Race/ethnicity is self-designated with codes as follows:

    B.Can = Black Canadian;

    B.Amer = Black American;

    Ital/Can = White, Italian Canadian;

    W.E. = Western European;

    E.E. = Eastern European;

    W.Can = White Canadian;

    W.Brit = White British;

    Pak/Can = Pakastani-Canadian;

    S.A. = South American;

    S.E.Asian = South East Asian;

    Port/Can = Portuguese-Canadian;

    B.Carr = Black Caribbean.

References

ABC Canada (1996). *Skills for Tomorrow: Basic Skills Project for the Union of Needle Trades, Industrial and Textile Employees*, Toronto, Canada.

Abella, I. (1975). "Oshawa 1937" in I. Abella (ed).) *On Strike: Six Key Labour Struggles in Canada: 1919-1949*. Toronto: James Lorimer.

ACTWU (1995). "The Changing Face of the Canadian Clothing Industry" in *ACTWEW Technology and Work Reorganization Project*. (Background Paper #2), Toronto, Canada.

Ainley, P. (1993). *Class and Skill: Changing Divisions of Knowledge and Labour*. London: Cassell.

Anderson, P. (1976). *Considerations on Western Marxism*. London: New Left Books.

Angus Reid Group. (2000). *Canadian Internet Access Continues to Grow, and Users Say the Net Has Had a Significant Impact on Their Lives*. (Ipsos-Reid Press Release) Ottawa, Canada.

Apple, M. (1990). *Ideology and Curriculum*. (2nd edition) London: Routledge.

Armstrong, P. and H. Armstrong (1995). *The Double Ghetto: Canadian Women and their Segregated Work*. (3rd edition) Toronto: McClelland & Stewart.

Arrowsmith, S. and C. Oikawa (2001). "Trends in Canadian Adult Education." in Statistics Canada *A Report on Adult Education and Training in Canada: Learning a Living*. Ottawa: Statistics Canada and Human Resources Development Canada.

Automotive Industry Sector Competitiveness Framework (1997). *Growth Prospects for the Industry*. Ottawa: Industry Canada. (*www.strategis.ic.gc.ca*)

Baca Zinn, M. and D. S. Eitzen (1993). *Diversity in Families*. (3rd edition) New York: Harper-Collins.

Barton, P. E. (2000). *What Jobs Require: Literacy, Education and Training, 1940-2006*. Princeton, NJ.: Educational Testing Service.

Becker, G. (1964). *Human Capital*. Chicago: University of Chicago Press.

Becker, H.S. (1970). *Sociological Work*. Chicago: Aldine.

Belanger, P. and P. Federighi (2000). *Unlocking People's Creative Forces: A Transnational Study of Adult Learning Policies*. Hamburg: UNESCO Institute for Education.

Berggren, C. (1993). "Designed for Learning: The Potential of Holistic Assembly as Contrasted to the Lean Line", Paper presented at *The Lean Workplace Conference*, Port Elgin, Ontario (Fall).

Bernstien, B. (1990). *The Structuring of Pedagogic Discourse*. London: Routledge.

—— (1996). *Pedagogy, Symbolic Control and Identity*. London: Taylor and Francis.

Betcherman, G., N. Leckie and K. McMullen (1997). *Developing Skills in the Canadian Workplace: The Results of the Ekos Workplace Training Survey*. Ottawa: Canadian Policy Research Networks.

Billett, S. (2001). "Co-participation: Affordances and Engagement at Work." *New Directions for Adult and Continuing Education*. 92, pp.63-72.

Blanck, G. (1993). "Vygotsky: The Man and his Cause." in L. C. Moll (ed.) *Vygotsky and Education: Instructional Implications and Applications of Sociohistorical Psychology*. Cambridge: Cambridge University Press.

Bohman, J. (1999). "Practical reason and cultural constraint: Agency in Bourdieu's Theory of Practice." in R. Shusterman (ed.) *Bourdieu: A Critical Reader*. Malden, MA.: Blackwell.

Bolaria, S. and P. Li (1988). *Racial Oppression in Canada*. Toronto: Garamond Press.

Boud, D. and D. Garrick (eds.) (1999). *Understanding Learning at Work*. London: Routledge.

Bourdieu, P. (1984). *Distinction: A Social Critique of the Judgement of Taste*. Cambridge: Harvard University Press.

—— (1991). *Language and Symbolic Power*. London: Polity.

—— (1994). *Academic Discourse: Linguistic Misunderstanding and Professorial Power*. Cambridge: Polity Press.

—— (1998). *Acts of Resistance:Against the Tyranny of the Market*. New York: W.W. Norton.

Bourdieu, P. and Contributors (1993). *La Misere du Monde*. Paris: Seuil. [English translation 1999]

Bourdieu, P. and J-C. Passeron (1977). *Reproduction in Education, Society and Culture*. London: Sage.

Bratton, J., J. Helms-Mills, T. Pyrch and P. H. Sawchuk (2003). *Workplace Learning: A Critical Introduction*. Toronto: Garamond.

Burkitt, I. (1991). *Social Selves: Theories of the Social Formation and Personality*. London: Sage.

Burn, D. (1997). "Recognizing Past Achievements." *Learning for the Workplace*. Nov. 17, 1997: L12-L14.

Burnett, J., D. Vincent and D. Mayall (1984). *The Autobiography of the Working Class: An Annotated Critical Bibliography*. Brighton: Harvester.

CAW Canada (1996). "How Workers Learn." *CAW Statement of Principles: Education*. (November). http://www.caw.ca/caw/cawedu.html.

—— (1997). *Canadian Auto Workers, Departments and Services*. (February). http://www.caw.ca/caw/webdepts.html.

Chaison, G. and J. Rose (1991). "Continental Divide: The Direction and Fate of North American Unions" *Advances in Industrial and Labor Relations*, 5, pp.169-205.

Cockburn, C. (1988). "The Gendering of Jobs: Workplace Relations and the Reproduction of Sex Segregation" in S. Walby (ed.) *Gender Segregation at Work*. Philadelphia, PA: Open University.

Collins, J. (1993). "Determination and Contradiction: An Appreciation and Critique of the Work of Pierre Bourdieu on Language and Education" in C. Calhoun (ed.) *Bourdieu: Critical Perspectives*. Chicago: University of Chicago Press.

Communications, Energy and Paperworkers Union of Canada (1994). *Working Families*. Toronto: CEP.

Corrigan, P. (1979). *Schooling for the Smash Street Kids*. London: Macmillan.

Courtney, S. (1992). *Why Adults Learn: Towards a Theory of Participation and Facilitating Learning*. London: Routledge.

Cuban, L. (1986). *Teachers and Machines: The Classroom Use of Technology Since 1920*. New York: Teachers' College Press.

Curtis, B., D.W. Livingstone and H. Smaller (1992). *Stacking the Deck: The Streaming of Working Class Kids in Ontario Schools*. Toronto: Our Schools/ Our Selves.

Dagg, A. (1994). "Virtual Corporations, Homework and Unions" *TECHnotes*. (July), No. 7, pp. 9-11.

Darrah, C. (1996). *Learning and Work: Explorations in Industrial Ethnography*. New York: Garland.

Das Gupta, T. (1996). *Racism and Paid Work*. Toronto: Garamond Press.

De Certeau, M. (1984). *The Practice of Everyday Life*. Berkeley: University of California Press.

Denning, M. (1998). *The Cultural Front*. London: Verso.

Dickenson, P. and G. Sciadas (1999). "Canadians Connected" *Canadian Economic Observer*. (Catalogue No. 11-010-XPB). Ottawa: Statistics Canada.

Dixon, N. (1992). "Organizational Learning: A Review of the Literature with Implications for HRD Professionals" *Human Resource Development Quarterly*, 3, pp.29-49.

Doeringer, P.B. and M. Piore (1971). *Internal Labor Markets and Manpower Analysis*. Lexington, MA: D.C. Heath.

Doray, P. and S. Arrowsmith (1997). "Patterns of Participation in Adult Education: Cross-national Comparisons" in P. Belanger and A. Tuijnman (eds.) *New Patterns of Adult Learning: A Six Country Comparative Study*. Oxford: Elsevier.

Dyer-Witherford, N. (1999). *Cyber-Marx: Cycles and Circuits of Struggle in High-Technology Capitalism*. Chicago: University of Illinois Press.

Edwards, R.C., M. Reich and D. Gordon (eds.) (1975). *Labour Market Segmentation*. Mass.: D.C. Heath & Co.

Elhammoumi, M. (2000). "Lost – or Merely Domesticated? The Boom in Socio-Historical Cultural Theory Emphasises Some Concepts, Overlooks Others" in S. Chaiklin (ed.) *The Theory and Practice of Cultural-Historical Psychology*. Aarhus: Aarhus University Press.

Engeström, Y. (1987). *Learning by Expanding: An Activity-theoretical Approach to Developmental Research*. Helsinki: Orienta-Konsultit.

—— (1992). *Interactive Expertise: Studies in Distributed Working Intelligence*. Helsinki: Department of Education, University of Helsinki.

—— (2000). "From Individual Action to Collective Activity and Back: Developmental Work Research as an Interventionist Methodology" in P. Luff, J. Hindmarsh and C. Heath (eds.) *Workplace Studies: Recovering Work Practice and Informing System Design*. New York: Cambridge University Press.

Engeström, Y., R. Engeström and M. Kärkkäinen (1995). "Polycontextuality and Boundary Crossing in Expert Cognition: Learning and Problem Solving in Complex Work Activities" *Learning and Instruction*, 5, pp.319-336.

Fals-Borda, O. and M.A. Rahman (eds.) (1991). *Action and Knowledge: Breaking the Monopoly with Participatory Action Research*. New York: Apex Press.

Fals-Borda, O. (1991). *Knowledge and Social Movements*. Santa Cruz, CA: Merrill Publishing.

Fashion Apparel Sector Campaign (1991). *Fashioning the Future: Building a Strategy for Competitiveness*. (Report on Phase II) Ottawa: Industry, Science and Technology Canada.

Foley, G. (1999). *Learning in Social Action: A Contribution to Understanding Informal Education*. London: Zed Books.

Forrester, K., J. Payne and K. Ward (1993). *Learning at Work: Final Report, Leeds Adult Learners at Work Project*. Leeds, UK: Department of Adult Continuing Education, University of Leeds.

—— (1995). *Workplace Learning: Perspectives on Education, Training and Work*. Aldershot, UK: Avebury.

Forrester, K. and C. Thorne (eds.) (1993). *Trade Unions and Social Research*. Aldershot, UK: Avebury.

Foucault, M. (1977). *Power/Knowledge: Selected Interviews and Other Writings*. New York: Pantheon.

Fowler, B. (1997). *Pierre Bourdieu and Cultural Theory*. London: Sage

Freeman, B. (1981). *1005: Political Life in a Union Local* Toronto: Lorimer.

Freire, P. (1970). *Pedagogy of the Oppressed*. New York: Continuum.

—— (1996). *Letters to Cristina: Reflections on my Life and Work*. New York: Routledge.

Freire, P. and D. Macedo (1987). *Literacy*. New York: Bergin and Garvey.

Freisen, G. (1994). "Adult Education and Union Education: Aspects of English Canadian Cultural History in the 20th Century" *Labour / Le Travail*, 34, pp.163-188.

Frenkel, S., M. Korczynski, K. Shire and M. Tam (1999). *On the Front Line: Organization of Work in the Information Economy*. Ithaca, NY: Cornell University Press.

Gannage, C. (1986). *Double Day, Double Bind: Women Garment Workers*. Toronto: Women's Press.

Garrick, J. (1999). *Informal Learning in the Workplace; Unmasking Human Resource Development*. London: Routledge.

Gaspasin, F and M. Yates (1997). "Organizing the Unorganized: Will Promises Become Practices?" *Monthly Review*. 49(3), pp.46-62.

Genovese, E. (1971). *Roll, Jordan, Roll: The World the Slaves Made*. New York: Vintage Books.

Gereluk, W. (2001). *Labour Education in Canada Today*. Athabasca, Canada: Athabasca University.

Ginden, S. (1995). *The Canadian Auto Workers: The Birth and Transformation of a Union*. Toronto: Lorimer.

Giroux, H. (1983). *Theory and Resistance in Education*. South Haddley, MA: Bergin and Garvey.

Goggan, P. (2003). "Report by Paul Goggan" (Certified Health and Safety Representative, Car Body, Hardware & South Stamping, Production and Skilled Trades and CAW Local 222 Executive Board Member) *The Oshaworker*, 62(1).

Gorman, R. (2002). "The Limits of 'Informal Learning': Adult Education Research and the Individualizing of Political Consciousness" in S. Mojab and W. McQueen (eds.) *Adult Education and the Contested Terrain of Public Policy. Canadian Association for Studies in Adult Education Conference Proceedings*. Toronto: Ontario Institute for Studies in Education of the University of Toronto.

Gramsci, A. (1971). *Selections from the Prison Notebooks*. New York: International Publishers.

Grant, M. (1992). "Industrial Relations in the Clothing Industry: Struggle for Survival" in R.P. Chaykowski and A. Verma (eds.) *Industrial Relations in Canadian Industry*. Toronto: Dryden.

Haddad, C. (1993). "Unions and Technological Change: Labor Research Strategies and Structures in the USA and Canada" in K. Forrester and C. Thorne (eds.) *Trade Unions and Social Research*. Aldershot, UK: Averbury.

Hamper, B. (1991). *Rivethead: Tales from the Assembly Line*. New York: Warner Books.

Handel, M. (2000). *Trends in Direct Measures of Job Skill Requirements*. (Working Paper No. 301, Jerome Levy Economics Institute - www.levy.org/does/wrkpap/papers).

Haraway, D. (1991). *Symians, Cyborgs and Women*. London: Free Association Books.

Hart, M. (1992). *Working and Educating for life: Feminist and International Perspectives in Adult Education*. New York: Routledge.

—— (1995). "Education and Social Change" in M. Newman (ed.) *Social Action and Emancipatory Learning*, (UTS Seminar Papers) Sydney, Australia: University of Technology Sydney, School of Education.

Hartsock, N. (1987). "The Feminist Standpoint: Developing the Ground for a Specifically Feminist Historical Materialism" in S. Harding and M. Hintikka (eds.) *Discovering Reality*. New York: D. Reidel Publishing Co.

Henry, F. (1994). *The Caribbean Diaspora in Toronto: Learning to Live with Racism*. Toronto: University of Toronto Press.

Herman, E. and N. Chomsky (1988). *Manufacturing Consent: The Political Economy of Mass Media*. New York: Pantheon.

Hewitt, P. (1993). *About Time: The Revolution in Work and Family Life*. London: River Oram Press.

hooks, b. (1984). *From Margin to Center*. Boston: South End Press.

Horton, M. and P. Freire (1990). *We Make the Road by Walking: Conversations on Education and Social Change*. Philadelphia, PA: Temple University Press.

Howell, Joseph T. (1973) *Hard Living on Clay Street*. New York: Anchor.

Human Resource Development Canada (1999) *Automotive Parts Industry Profile*. Ottawa: Government of Canada.

—— (2002). *Highlights: Auto Assembly*. Ottawa: Government of Canada.

ILGWU (1993). *ILGWU 1993 Homeworkers' Study: An Investigation into Wages and Working Conditions of Chinese-speaking Homeworkers in Metropolitan Toronto. Summary of Study Findings*. (Jan Buroway and Fanny Yuen) Toronto: ILGWU.

—— (1995). *Designing the Future for Garment Workers*. (TARP Project) Toronto: ILGWU.

ILGWU/UNITE (1995). *The Workplace Issues Survey*. Toronto: ILGWU/UNITE.

Ilyenkov, E. V. (1982). *The Dialectics of the Abstract and the Concrete in Marx's Capital*. Moscow: Progress Publishers.

Industry Canada (1997a). *Canadian Chemical Industry Statistical Handbook*. Ottawa: Government of Canada.

—— (1997b). *Automotive Industry Sector Competitive Framework*. Ottawa: Government of Canada.

—— (2002). *Canadian Industry Statistics: Clothing Manufacturing* . Ottawa: Government of Canada.

Information Highway Advisory Council (1995). *Connecting Community Content: The Challenge of the Information Highway*. Ottawa: Minister of Supply and Services, Government of Canada.

International Labour Organization (1995). *Recent Developments in the Clothing Industry*. (Sectoral Activities Programme) Geneva: ILO.

Jackson, B. and D. Marsden (1986). *Education and the Working Class*. London: Ark Publishing.

Joseph, S. (1983). "Working-Class Women's Networks in a Sectarian State: A Political Paradox" *American Ethnologist*, 10, pp. 1-22.

Joyce, P. (1991). *Visions of the People: Industrial England and the Question of Class. 1848-1914*. Cambridge: Cambridge University Press.

Kane, E. and D. Marden (1988). "The Future of Trade Unionism in Industrial Market Economies" *Labour and Society*, 13(2), pp.107-124.

Keenan, G. (1996). "GM Canada Sets Profit Mark: New High for Canadian Companies of $1.391 Billion Helps U.S. Parent Post Record" *The Globe and Mail* (Toronto, January 31).

Kempston Darkes, M. (1995). "Remarks to the 1995 Automotive News World Congress." Detroit, Michigan, January 9, 1995.

Knowles, M. (1970). *The Modern Practice of Adult Education: Andragogy Versus Pedagogy*. Chicago: Follett.

—— (1975). *Self directed Learning*. New York: Associated Press.

—— (1977). *A History of the Adult Education Movement in the United States*. New York: Krieger.

Kohn, M. and C. Schooler (1983). *Work and Personality: An Inquiry in to the Impact of Social Stratification*. Norwood, NJ: Ablex Publishing.

Kok, J. (2002). "Preface" in J. Kok (ed.) *Rebellious Families: Household Strategies and Collective Action in the Nineteenth and Twentieth Centuries*. New York: Berghahn Books.

Krahn, H. and G. Lowe (2002). *Work, Industry and Canadian Society*. Toronto: Thompson.

Kumar, P. (1993). "The Canadian Experience: Establishing and Strengthening Links between Trade Unions and Researchers" in K. Forrester and C. Thorne (eds.) *Trade Unions and Social Research*. Aldershot, UK: Averbury.

Kusterer, K. (1978/1982). *Know How on the Job: The Important Working Knowledge of "Unskilled" Workers*. Boulder: Westview Press.

Lareau, A. (1989). *Home Advantage: Social Class and Parental Inntervention in Elementary Education*. New York: Falmer.

Lasch, S. and J. Urry. (1994). *Economies of Signs and Space*. London: Sage.

Lavoie, M. and R. Roy. (1998). *Employment in the Knowledge-Based Economy: A Growth Accounting Exercise for Canada*. (Research Paper R-98-8E) Ottawa: Applied Research Branch, Human Resources Development Canada.

Lave, J. (1993). "The Practice of Learning" in S. Chaiklin (ed.) *Understanding Practice: Perspectives on Activity and Context*. Cambridge: Cambridge University Press.

Lazonick, W. (1991). *Business Organization and the Myth of the Market Economy*. New York: Cambridge University Press.

Leckie, N. (1996). *On Skill Requirements Trends in Canada, 1971-1991*. Research Report for Human Resources Development Canada and Canadian Policy Research Networks.

Leont'ev, A.N. (1978). *Activity, Consciousness and Personality*. Englewood Cliffs: Prentice-Hall.

—— (1981). *Problems of the Development of the Mind*. Moscow: Progress.

Lewchuk, W., B. Roberts and C. McDonald (1996). *The CAW Working Conditions Study: Benchmarking Auto Assembly Plants*. Willowdale: CAW Canada.

Lipsig-Mummé, C. (1987). "Organizing Women in the Garment Trades: Homework and the 1983 Garment Strike in Canada" *Studies in Political Economy*, 22 (Spring).

Livingstone, D.W. (ed.) (1987). *Critical Pedagogy and Cultural Power*. Haddley, MA: Bergin and Garvey.

—— (1997a) "The Limits of Human Capital Theory: Expanding Knowledge, Informal Learning and Underemployment" *Policy Opinions*. 18(6).

—— (1997b). "Computer Literacy, the 'Knowledge Economy' and Information Control: Micro Myths and Macro Choices" in M. Moll (ed.) *Tech High: Globalization and the Future of Canadian Education*. Ottawa: Canadian Centre for Policy Alternatives.

—— (1999a). *The Education-Jobs Gap: Underemployment or Economic Democracy*. Toronto: Garamond Press (second edition 2003).

—— (1999b). "Exploring the Icebergs of Adult Learning: Findings of the First Canadian Survey of Informal Learning Practices" *Canadian Journal for the Study of Adult Education*, 13(2), pp.49-72.

—— (2001). *Adults' Informal Learning: Definitions, Findings, Gaps and Future Research*. Position paper for the Advisory Panel of Experts on Adult Learning, Applied Research Branch, Human Resources Development Canada.

—— (2002). *Working and Learning in the Information Age: A Profile of Canadians*. Ottawa: Canadian Policy Research Networks.

Livingstone, D.W., D. Hart and L. Davie. (1999). *Public Attitudes Toward Education in Ontario 1998: Twelfth OISE/UT Survey*. Toronto: University of Toronto Press.

—— (2001). *Public Attitudes Toward Education in Ontario 2000: Thirteenth OISE/UT Survey*. Toronto: University of Toronto Press.

Livingstone, D.W. and J.M. Mangan (1993). "Class, Gender, and Expanded Class Consciousness in Steeltown" *Research in Social Movements, Conflicts and Change*, 15, pp. 55-82.

Livingstone, D.W. and J.M. Mangan (eds.) (1996). *Recast Dreams: Class and Gender Consciousness in Steeltown*. Toronto: Garamond Press.

Livingstone, D.W, M. Raykov and S. Stowe. (2002). *Interest in and Factors Related to Participation in Adult Education and Informal Learning*. (Research Paper R-01-9-3E) Hull: Applied Research Branch, Human Resources Development Canada.

Livingstone, D.W. and R. Roth (1997). *Building a Social Movement Community: Oshawa Autoworkers*. Paper presented at annual meeting of the Canadian Association for the Study of Education, St. John's, Nfld, June 9-11.

Livingstone, D.W. and D.E. Smith (forthcoming). *The Decline of Steelwork*.

Livingstone, D.W. and S. Stowe (2003). "Class and Learning in Canada: Intergenerational Patterns of Inequality" in A. Scott and J. Freeman-Moir (eds.) *Yesterday's Dreams: International and Critical Perspectives on Education and Social Class*. Auckland: Canterbury University Press.

London, S., E. Tarr and J. Wilson (eds) (1990). *The Education of the American Working-Class*. New York: Greenwood.

Lowe, G. (1992). *Human Resource Challenges of Education, Computers and Retirement.* Ottawa: Statistics Canada.

—— (1996). *The Use of Computers in the Canadian Workplace.* Paper prepared for Information Technology Innovation, Industry Canada, March 3ʳᵈ.

—— (2000). *The Quality of Work: A People-Centred Agenda.* New York: Oxford University Press.

Lukács, G. (1971). *History and Class Consciousness.* Cambridge, MA: MIT Press.

Luke, C. and J. Gore (eds.) (1992). *Feminisms and Critical Pedagogy.* New York: Routledge.

Luria, A.R. (1976). *Cognitive Development, its cultural and social foundations.* (edited by Michael Cole) Cambridge: Harvard University Press.

Luttrell, W. (1997). *Schoolsmart and Motherswise: Working-class Women's Identity and Schoooling.* New York: Routledge.

Luxton, M. and J. Corman (2000). *Getting by in Hard Times: Gender and Class Relations in Steeltown.* Toronto: University of Toronto Press.

Luxton, M. (1980). *More than a Labour of Love: Three Generations of Women's Work in the Home.* Toronto: The Women's Press.

MacLeod, J. (1995). *Ain't No Making It: Aspirations and Attainment in a Low-Income Neighbourhood.* Colorado: Westview Press.

Manley, J. (1986). "Communists and Auto Workers: The Struggle for Industrial Unionism in the Canadian Auto Industry, 1925-36" *Labour/Le Travail,* (Spring).

Marsick, V. and K. Watkins (1990). *Informal and Incidental Learning in the Workplace.* London: Routledge.

Martin, D. (1995). *Thinking Union.* Toronto: Between the Lines Press.

Martinez, L. and S. Eston (1992). "Human Resource Management and Trade Union Responses: Bring the Politics of the Workplace into the Debate" in P. Blyton and P. Turnbull (eds.) *Reassessing Human Resource Management.* London: Sage.

McIlroy J. and S. Westwood (eds.) (1993). *Border Country: Raymond Williams in Adult Education.* Leicester: National Institute of Adult Continuing Education.

Mead, G. H. (1934). *Mind, Self, and Society.* Chicago: University of Chicago Press.

Mezirow, J. (1991). *Transformative Dimensions of Adult Learning.* San Francisco: Jossey-Bass.

—— (1994). "Understanding Transformative Theory" *Adult Education Quarterly,* 44(4), pp.222-244.

—— (1996). "Contemporary Paradigms of Learning" *Adult Education Quarterly,* 46(3), pp.158-173.

Milgram, S. (1974). *Obedience to Authority.* New York: Harper and Row.

Milkman, R. (1997). *Farewell to the Factory: Auto Workers in the Late Twentieth Century.* Berkeley: University of California Press.

National Research Council of the United States (1991). *Work and Family: Policies for a Changing Work Force.* Washington: National Academy Press.

Mishel, L. and P. Voos (eds) (1992). *Unions and Economic Competitiveness.* New York: M.E. Sharpe Inc.

Negt, O. and A. Kluge (1993). *Public Sphere and Experience: Toward an Analysis of the Bourgeois and Proletarian Public Sphere.* Minneapolis: University of Minnesota Press.

Newman, F. and L. Holzman (1993). *Lev Vygotsky: Revolutionary Scientist.* London: Routledge.

Newman, M. (1993). *The Third Contract: Theory and Practice in Trade Union Training.* Sydney: Stewart Victor.

—— (1994). *Defining the Enemy: Adult Education in Social Action.* Sydney: Stewart Victor.

Ng, R. (1988). *The Politics of Community Services*. Toronto: Garamond Press.

—— (1994). *Apparel Textile Action Committee, Final Report*. Toronto: OISE/UT.

—— (2002). "Globalization and Garment Workers in Canada: Implications for Social Policy" (Working Paper Number 3) *Changing Work, Changing Lives: Mapping the Canadian Garment Industry*. Department of Adult Education and Counselling Psychology, OISE/UT, Canada.

Ng, R. and A. Estable (1987). "Immigrant Women in the Labour Force: An Overview of Present Knowledge and Research Gaps" *Resources for Feminist Research*. 16(1).

OECD. (1998). "Lifelong Learning: A Monitoring Framework and Trends in Participation" in *Education Policy Analysis*. (Centre for Educational Research and Innovation) Paris: OECD, pp.7-24.

Ollman, B. (1993). *Dialectical Investigations*. London: Routledge.

Parenti, M. (1996). *Dirty Truths*. San Francisco: City Lights Books.

Parker, M. and J. Slaughter (1994). *Working Smart: A Union Guide to Participation Programs and Reengineering*. Detroit: Labour Notes.

Pearl, A. (1997). "Democratic Education as an Alternative to Deficit Thinking" in R. Valencia (ed.) *The Evolution of Deficit Thinking*. London: Falmer Press.

Peirce, C. S. (1931-1935) *Collected Papers of Charles Sanders Peirce*. (edited by C. Hortshorne & P. Weiss) Cambridge, Mass.: Harvard University Press.

Percy, K., D. Burton and A. Withnall (1994). *Self-Directed Learning among Adults: The Challenges for Continuing Educators*. Lancaster, UK: University of Lancaster.

Poonwasie, D. and A. Poonwasie (eds.) (2001). *Fundamentals of Adult Education: Issues and Practices for Lifelong Learning*. Toronto: Thompson.

Prakesh, M.S. and G. Esteva. (1998). *Escaping Education: Living as Learning within Grassroots Cultures*. New York: Peter Lang.

Reason, P. and H. Bradbury (eds.) (2001). *Handbook of Action Research, Participative Inquiry and Practice*. London: Sage.

Reddick, A., C. Boucher and M. Groseilliers. (2000). *The Digital Divide: The Information Highway in Canada*. Ottawa: Public Interest Advocacy Centre.

Reich, R. (1991). *The Work of Nations*. New York: Vintage Press.

Rinehart, J. (1997). *The Tyranny of Work: Alienation and the Labour Process*. Toronto: Harcourt Brace Jovanovich Canada.

Rinehart, J., C. Huxley and D. Robertson (1997). *Just Another Car Factory? Lean Production and Its Discontents*. Ithaca: ILR Press.

Robertson, H. (1995). *Driving Force: The McLaughlin Family and the Age of the Car*. Toronto: McClelland & Stewart.

Rose, N. (1990). *Governing the Soul: The Shaping of the Private Self*. London: Routledge.

Roth, R. (1997). *Kitchen Economics for the Family@: Paid Education Leave in the Canadian Region of the United Auto Workers*. M.A. Thesis, University of Toronto.

—— (forthcoming). *Oshawa Autoworkers: Social Integration and Oppositional Class Consciousness*. Ph.D. dissertation, University of Toronto.

Rowlands, S. (2000). "Turning Vygotsky on His Head: Vygotsky's 'Scientifically Based Method' and the Socioculturalist's 'Social Other'" *Science and Education*. 9, pp.537-575.

Rubin, L. (1976). *Worlds of Pain: Life in a Working-Class Family*. New York: Basic Books.

—— (1994). *Families on the Fault Line: America's Working Class Speaks out about Family, the Economy, Race and Ethnicity*. New York: Harper-Collins.

Ruiz, R. (1984). "Orientations in Language Planning" in S. McKay and S. Wong (eds.) *Language Diversity: Problem or Resource*. New York: Newbury House Publisher.

Sadovnik, A. (ed.) (1995). *Knowledge and Pedagogy: The Sociology of Basil Bernstein*. Norwood, NJ: Ablex.

Said, E. (1993). *Culture and Imperialism*. Chatto and Windus: London.

Sargent, L. (ed.) (1981). *The Unhappy Marriage of Marxism and Feminism: A Debate on Class and Patriarchy*. London: Pluto Press.

Sawchuk, P.H. (1996). *Working Class Informal Computer Literacy*. MA Thesis, University of Toronto.

—— (1998). "'This Would Scare the Hell out of me if I were a HR Manager': Workers Making Sense of PLAR" in the *1998 Canadian Association for Studies in Adult Education Conference Proceedings*. University of Ottawa, Canada.

—— (2000). "Building Traditions of Inquiry and Transforming Labour-Academic Collaboration at the Union Local: Case Studies of Workers' Research and Education" *Labour / Le Travail*, 45(Spring), pp.199-216.

—— (2003a). *Adult Learning and Technology in Working-Class Life*. New York: Cambridge University Press.

—— (2003b). "Informal Learning as a Speech-Exchange System: Implications for Knowledge Production, Power and Social Transformation" *Discourse and Society*, 14(3), pp.291-307.

—— (in press) "The 'Unionization Effect' amongst Adult Computer Learners" *British Journal of Sociology of Education*, 24(5).

Schied, F., V. Carter, J. Preston and S. Howell (1997). "The HRD Factory: An Historical Inquiry into the Production of Control in the Workplace" in M. Zukas (ed.) *Crossing Boarders, Breaking Boundaries: Proceedings of the Standing Committee on University Teaching, Research and Education of Adults Conference*. Birkbeck College, University of London, U.K.

Schroeder, A. (1984). "Autobiography of Allen Eugene Schroeder." *Schroeder Papers*. (MG31 B43), National Archives of Canada.

Schön, D., B. Sanyal and W. J. Mitchell (eds.) (1999). *High Technology and Low-Income Communities: Prospects for the postive use of advanced information technology*. Cambridge, MA: MIT Press.

Sciadis, G. (2002). *The Digital Divide in Canada*. (Catalogue No. 56F009X1E) Ottawa: Statisitics Canada.

Seccombe, W. and D.W. Livingstone (1999). *Down to Earth People: Beyond Class Reductionism and Postmodernism*. Toronto: Garamond Press.

Selman, G. and P. Dampier (1991). *The Foundations of Adult Education in Canada*. Toronto: Thompson Educational Publishing.

Senge, P. (1990). *The Fifth Discipline: The Art and Practice of the Learning Organization*. New York: Doubleday Currency.

Sennett, R. and J. Cobb (1972) *The Hidden Injuries of Class*. New York: Vintage.

Sharp R., M. Hartwig and J. O'Leary (1989). "Independent Working Class Education: A Repressed Historical Alternative" *Discourse*. 10(2), pp. 1-26.

Sheikh, Y. (1999). *Has the Clothing Industry Adapted to the Changing Economic Environment?* (Publication #34-252) Ottawa: Statistics Canada.

Simon, B. (ed.) (1990). *The Search for Enlightenment: The Working Class and Adult Education in the Twentieth Century*. London: Lawrence and Wishart.

Sobel, D. and J. Stephen (1996). *Fashioning a Future: Prospects for the Fashion District in the City of Toronto*. Toronto: New View Research.

Sochor, Z. (1988). *Revolution and Culture: The Bogdanov-Lenin Controversy*. Ithaca: Cornell University Press.

Speech from the Throne to Open the Second Session of the Tthirty-Sixth Parliament of Canada. (http://www.pco-bcp.gc.ca/sft-ddt/doc/fulltext).

Spencer, B. (1994). "Educating Union Canada" *Canadian Journal for the Study of Adult Education.*8(2), pp.45-64.

—— (ed.) (2002). *Unions and Learning in a Global Economy: International and Comparative Perspectives.* Toronto: Thompson.

Spender, D. (1980). *Man Made Language.* London: Pandora.

Statistics Canada (1998). "Education, Mobility and Migration" *The Daily.* (April 14) Ottawa: Statistics Canada.

—— (1999a). *Adult Education and Training Survey 1998.* (Catalogue No. 81C0045) Ottawa: Statistics Canada.

—— (1999b). "Household Spending, Dwelling Characteristics and Household facilities" *The Daily* (December 13th).

—— (1999c). "Supplementary Measures of Unemployment" *Labour Force Update* 3, pp.1-23.

—— (2000). "Education Indicators," *The Daily* (February 21st).

—— (2001). "For the Sixth Consecutive Year, the Highest Median Income was Found in Oshawa ($62,500) and Windsor ($62,400)" *The Daily* (August 10th).

—— (2002). "Household Internet Use Survey" *The Daily* (July 25th).

Steedman, M. (1986). "Skill and Gender in the Canadian Clothing Industry, 1890-1940" in C. Heron and R. Storey (eds.) *On the Job: Confronting the Labour Process in Canada.* Kingston: McGill-Queen's University Press.

Sugiman, P. (1994). *Labour's Dilemma: The Gender Politics of Auto Workers in Canada, 1937-1979.* Toronto: University of Toronto Press.

Swartz, D. (1997). *Culture and Power: The Sociology of Pierre Bourdieu.* Chicago: University of Chicago Press.

Tandan, N. (1969). *Underutilization of Manpower in Canada.* (Special Labour Force Studies No. 8) Ottawa: Dominion Bureau of Statistics.

Thompson, E. P. (1968). *Education and Experience.* Leeds: Leeds University Press.

Taylor, J. (2001). *Union Learning: Canadian Labour Education in the Twentieth Century.* Toronto: Thompson.

Tilly, L. and C. Tilly (eds.) (1981). *Class Conflict and Collection Action.* Beverly Hills, CA: Sage.

Tough, A. (1967) *Why adults learn: A study of the major reasons for beginning and continuing a learning project.* Toronto: OISE.

—— (1978). "Major learning efforts: Recent research and future directions," *Adult Education,* 28, pp.250 - 63.

—— (1979). *The Adult's Learning Projects: A Fresh Approach to Theory and Practice in Adult Learning.* Toronto: OISE Press.

Traill, B. (1997). *CAW Strike 1996.* [video recording] Toronto: CAW.

Turner, G.R. and R. P. Hadfield (eds.) (1994). *Total Quality Management in the Chemical Industry: Strategies for Success.* Cambridge, UK: Royal Society of Chemistry.

Turner, H.A. (1962). *Trade Union Growth, Structure and Policy,* London: Allen and Unwin.

UNITE (1995). "The First Truly United Project" *UNITE's National Canadian Newsletter,* December, p.7.

University and College Labor Education Association (with the American Association of Community and Junior Colleges, AFL-CIO and UAW) (1977). "Joint Statement on Effective Cooperation between Organized Labor and Higher Education" *Labor Studies Journal,* 1(3), pp.291-296.

Valencia, R.R. (1997). *The Evolution of Deficit Thinking: Educational Thought and Practice*. Washington, DC: Falmer Press.

Van Alphen, Tony. (1996). "GM Strike Finally over but Big Issue Unsettled" *The Toronto Star*, [Toronto, March 23rd].

Van der Linden, M. (1994). "Households and Labour Movements" *Economic and Social History*, 6, pp.129-144.

—— (2002). "Conclusion" in J. Kok (ed.) *Rebellious Families: Household Strategies and Collective Action in the Nineteenth and Twentieth Centuries*. New York: Berghahn Books.

Vygotsky, L.S. (1978). *Mind in Society: The Development of Higher Psychological Processes*. (edited by M. Cole) Cambridge: Harvard University Press.

—— (1994). *The Vygotsky Reader*. (edited by R. Van der Veer and J. Valsiner) London: Blackwell.

Waring, M. (1988). *If Women Counted: A New Feminist Economics*. San Francisco: Harper and Row.

Weber, T. (2003). "GM Gears up Earnings" *Globe and Mail* [Toronto, January 17th].

Weis, M. (2001). "Online America" *American Demographics*. March (*www.demographics.com*).

Wells, D. (1987). *Empty Promises: Quality of Working Life Programs and the Labour Movement*. New York: Monthly Review Press.

Welton, M. (1995). *In Defense of the Lifeworld: Cricial Perspectives on Adult Learning*. Albany, N.Y.: SUNY Press.

Wertsch, J. V. (1991). *Voices of the Mind: A Sociocultural Approach to Mediated Action*. Cambridge: Harvard University Press.

White, B. (1987). *Hard Bargains: My Life on the Line*. Toronto: McLelland and Stewart.

Williams, R. (1975). "You're a Marxist, Aren't you?' in B. Parekh (ed.) *The Concept of Socialism*. London: Croom Helm.

—— (1977). *Marxism and Literature*. London: Oxford University Press.

Willinsky, J. (1998). *Learning to Divide the World: Education at Empire's End*. Minneapolis: University of Minneapolis Press.

Willis, P. (1977/1981). *Learning to Labour: How Working Class Kids Get Working Class Jobs*. Farnborough: Saxon House.

Womack, J., D. Jones and D. Roos (1990). *The Machine that Changed the World*. Toronto: Collier Macmillan.

Worthen, H. (2001). "Studying the Workplace: Considering the Usefulness of Activity Theory" *XMCA Paper Archive* (*www.*communication.ucsd.edu/MCA/Paper).

Wright, E.O. (1996). *Classes*. London: Verso.

Yates, C. (1993). *From Plant to Politics: The Autoworkers Union in Postwar Canada*. Philadelphia: Temple.

Zandy, J. (ed.) (2001). *What We Hold in Common: An Introduction to Working-Class Studies*. New York: Feminist Press at the City University of New York.

Index